Fibre-reinforced Concretes for High-performance Structures

Fibre-reinforced Concretes for High-performance Structures

Building a more sustainable future

Dr Andreas Lampropoulos

Published by Emerald Publishing Limited, Floor 5, Northspring, 21-23 Wellington Street, Leeds LS1 4DL.

ICE Publishing is an imprint of Emerald Publishing Limited

Other ICE Publishing titles:
ICE Handbook of Concrete Durability, Second edition
Edited by Marios Soutsos. ISBN 978-0-7277-6375-4
Appraisal and Repair of Existing Concrete Structures
Yue Choong Kog. ISBN 978-0-7277-6693-9
Concise Guide to Reinforced Concrete Design to Eurocode 2
Patrick Purcell. ISBN 978-0-7277-6572-7

A catalogue record for this book is available from the British Library

ISBN 978-0-7277-6556-7

Cover photo: Jean Bouin stadium in Paris. Brian Ormerod/Alamy Stock Photo

Commissioning Editor: Viktoria Hartl-Vida
Assistant Editor: Cathy Sellars
Production Editor: Sirli Manitski

Typeset by Manila Typesetting Company
Index created by Martin Brooks

Dedication

To Rania, Maria & Panagiota
and to my parents Panagioti & Rania

Contents

Preface

Concrete is one of the most commonly used materials in the construction industry. This is attributed to its high compressive strength and relatively low cost. However, there are drawbacks with the use of concrete which are mostly linked to the fact that cement – one of its key ingredients – releases enormous amounts of carbon dioxide during its production phase (1 tonne of cement emits almost 0.8 tonnes of carbon dioxide). Worldwide, there is an ever-increasing pressure for countries, governments and companies to become 'greener' by decreasing their carbon dioxide emissions, which are the main cause of global warming. In addition to environmental aspects, enhanced structural performance is always a key priority and, therefore, there is a need for the extensive use of new forms of sustainable and high-performance concretes in construction.

This book deals with recent developments in the production of types of concrete with enhanced structural performance and sustainability. In the last decade, there have been significant advances in the development of novel types of concrete for structural applications. Research studies have focused mostly on the enhancement of the structural performance and durability, although substantial efforts have been made to reduce the environmental impact linked to concrete production. Despite the significant amount of research and the very encouraging results in this area, the construction industry has been very slow to adopt new concrete types, the reason for which is mostly linked to the limited knowledge of practitioners in this field and the lack of available standards. This book focuses on addressing this gap by presenting key information about the development, performance and design of three main materials: (*a*) fibre-reinforced concrete (FRC), (*b*) ultra-high-performance fibre-reinforced concrete (UHPFRC) and (*c*) geopolymer concrete (GC) (including fibre-reinforced GC). Recent research findings are collected and critically analysed to highlight the key parameters for the development and production process and to characterise the properties of the materials, with a focus on the mechanical properties and durability characteristics. Design aspects are also covered using both research outcomes and available guidelines/code provisions. Critical evaluation of the mechanical properties of all of the examined materials are presented, in addition to environmental and economic considerations, which makes this book a key resource for the selection of suitable types of concrete for structural applications. It is expected that this publication will facilitate the quicker adoption and the extensive use of these novel types of concrete in industry.

In closing, the author would like to express his appreciation and sincere thanks to Professor Kypros Pilakoutas for providing valuable information about one of the case studies presented in the book.

About the author

Dr Andreas Lampropoulos is a Principal Lecturer in Civil Engineering at the University of Brighton. He obtained his Diploma (2003), MSc (2005) and PhD (2010) degrees in Civil Engineering (Structural Division) from the University of Patras in Greece.

His main research agenda spans the areas of novel construction materials and seismic strengthening/retrofitting of existing structures. His research interests are focused on a wide range of cementitious materials such as ultra-high-performance fibre-reinforced concrete (UHPFRC), steel-fibre-reinforced concrete (SFRC) and cementitious materials reinforced with nanoparticles, and he is also working on the development and application of cement-free concretes. He has conducted extensive experimental and numerical work on the development of novel strengthening techniques for the structural upgrade of reinforced concrete (RC) and unreinforced masonry (URM) structures.

Dr Lampropoulos currently serves as the Chair of the International Association for Bridge and Structural Engineering (IABSE) Task Group 1.1 'Improving Seismic Resilience of Reinforced Concrete Structures' and Task Group 5.5 'Conservation and Seismic Strengthening/Retrofitting of Existing Unreinforced Masonry Structures', and he is an active member of various task groups of IABSE and of the International Federation for Structural Concrete (fib).

He has had more than 80 works published in international research journals, books and conference proceedings, and he is an editorial board member of international journals in this field. In addition, he is a reviewer for more than 35 world-leading international journals, a member of the scientific committee of more than 20 international conferences, and he is an invited reviewer for research proposals for national centres of various countries.

Lampropoulos A
ISBN 978-0-7277-6556-7
https://doi.org/10.1680/fchs.65567.001

Chapter 1
Fibre-reinforced concrete (FRC)

1.1. Introduction and historical developments

Concrete is one of the most commonly used construction materials. It is characterised by superior compressive strength, relatively low cost and ease of application. The main drawbacks are linked to the relatively low tensile performance and potential for cracking along with the subsequent durability issues. In addition, the extensive use of concrete in construction gives rise to significant carbon dioxide emissions which has catastrophic environmental consequences.

An enormous amount of research has been conducted in the last few decades on the development of novel high-performance and environmentally friendly types of concrete that are reinforced with different types of fibres.

In the case of fibre-reinforced concrete (FRC) the concept is quite similar to that of conventionally reinforced concrete (RC) with steel bars. The cementitious matrix, which can be either concrete or mortar, is normally characterised by high compressive strength and has numerous additional benefits such as enhanced fire resistance, sound insulation and protection against vibration. However, the weak tensile strength characteristics make it prone to tensile failures and cracks, and therefore the addition of steel reinforcement is essential for the enhancement of the structural performance and durability of concrete structures.

A key parameter for the performance of RC elements is the bond between the steel reinforcement and the cementitious matrix, which is essential for the interaction of the two materials and the composite action. The steel reinforcement can be either in the form of continuous steel bars or, in case of fibre reinforcement, the reinforcement is considered discontinuous. Continuous reinforcement has been extensively used in many different types of concrete since 1855, when thin elements of mortar reinforced with steel mesh were used to construct a boat. The use of RC in small-scale structures was also reported around the same period and, since then, further developments have been made, such as the use of prestressed concrete (Figure 1.1).

The concept of FRC is not new. In fact, fibres have been used since about 1500 BC. At that time horsehair was used in mortar and straw in mud-bricks [1]. Systematic studies recording the addition of fibres for the enhancement of the tensile and bending resistance of concrete were made from 1874 onwards (Figure 1.2). In 1874, A. Berard patented for the first time the use of irregular waste iron pieces for the strengthening of 'artificial stones' which were made of concrete. In the 1900s, asbestos fibres were used in concrete, but their use was discontinued owing to the high risk of health issues caused by the use of asbestos. After that, the use of

Figure 1.1 Overview of the history of concrete with steel reinforcement [2]

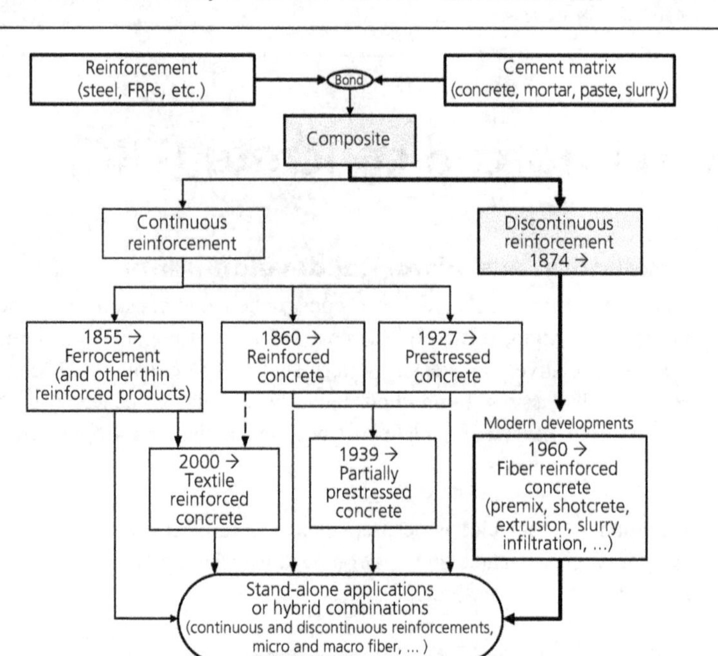

Figure 1.2 FRC developments and overview of the main milestones [2]

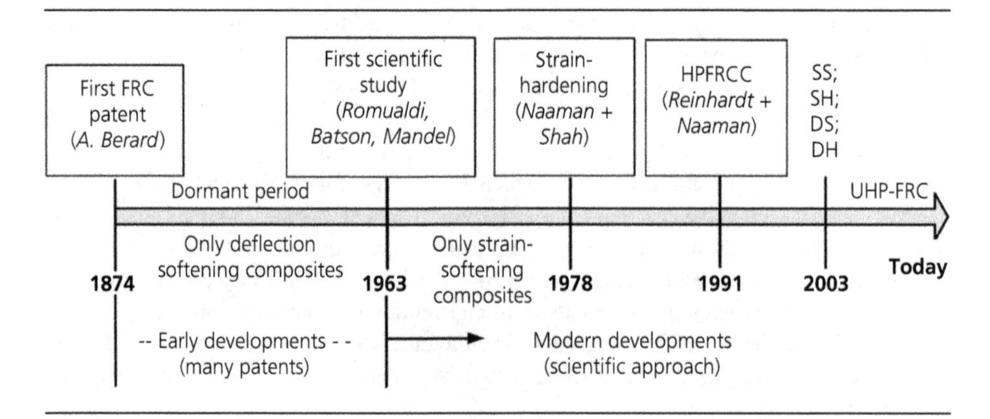

various types of fibre was explored, with steel, glass and synthetic ones being the most popular. In the early 1960s, the first scientific papers were published by Romualdi, Batson and Mandel which offered an in-depth understanding of the mechanics of crack arrest and the role of the orientation of the fibres. Then, and until 1978, the behaviour of the examined FRC types was characterised by strain-softening characteristics, whereas, after that period, developments were made on composites with strain-hardening behaviour. High-performance and ultra-high-performance fibre-reinforced composites were first developed in 1991 and 2003, respectively [2].

Modern developments in the types of FRC and the continuous improvement of the mechanical characteristics are attributed to a large extent to the developments in the types of fibres that were introduced in the mid-1980s [3]. The new fibre types, in addition to the development of new types of admixtures and production methods, have led to beneficial effects on fibre concentration and orientation with subsequent enhancement of the mechanical properties [3].

1.2. Materials selection

The selection of the materials and the mix design are the key parameters for the performance of FRC. In this section, an overview of the most commonly used cement, aggregates and fibre types are presented, in addition to the mix design processes.

1.2.1 Cement

In the case of FRC, the concept is quite similár to that used for conventional concrete and there are not specific requirements for the type of cement that should be used. For 'conventional' FRC, ordinary Portland cement is normally used unless there are special requirements in terms of structural performance (for example, high-performance concrete (HPC) and ultra-high-performance concrete (UHPC)) where high-strength cement types should be used. The effect of cement type is directly linked to the compressive strength of the concrete and subsequently affects the mechanical characteristics of FRC.

Simões *et al.* (2017) [4] investigated the effect of different types of cement on the characteristics of FRC. CEM II/B-L 32.5 R, CEM II/A-L 42.5 R and CEM I 52.5 R were adopted for 20, 60 and 100 MPa compressive strength mixtures and the results show that an increase in concrete compressive strength leads to an increase in the number of original fibres that fail by tensile strength, instead of slipping and debonding of the concrete matrix. This is attributed to the increment of the interface bond strength and to the enhancement of the strength of the adjoining matrix in the case of hooked-end fibres [4].

1.2.2 Aggregates

The size of the aggregates in FRC is a key parameter for the mechanical performance and durability of the material and this is attributed to the synergistic effect at the interfaces between the steel fibres and the aggregates.

Seleem *et al.* (2020) [5] investigated the effect of various aggregate dimensions with maximum aggregate size 10 mm, 20 mm, 25 mm and 40 mm and two different percentages of steel fibres, 0% and 1%. Compressive, indirect tensile and flexural strength tests were conducted and the results show that all of these mechanical characteristics were increased as the maximum aggregate size was increased, and increments were also observed with the addition of steel fibres. There are various other studies where the increment of the compressive strength and, subsequently, the tensile and flexural strength characteristics were increased for higher maximum aggregate size [6–10].

However, the effect of the aggregate size on the flexural energy in the post-peak region may be detrimental and this depends on the aggregate size and on the fibre geometry. According to Seleem *et al.* (2020) [5], the flexural load in the post-peak region of the load–deflection results was reduced as the maximum aggregate size was increased, which is linked to the reduced bond between the fibres and the cementitious matrix. On the other hand, Ulas *et al.* (2017) [6] observed a small increment in the flexural toughness around 10% when the maximum aggregate size was increased from 16 mm to 31.5 mm. The fibre length is also an important parameter for the fibre-to-cementitious-matrix bond [8]. Specimens with shorter fibres experience higher ultimate direct tensile strength values. However, the ultimate strain is reduced and there is more likely to be reduced post-cracking performance than in specimens with longer fibres, in which enhanced bonding between the fibres and the cementitious matrix is achieved [8].

According to the study of Han *et al.* 2019 [9], the indirect splitting tensile strength of steel-fibre-reinforced concrete (SFRC) gradually increases with increase in the length of the steel fibre. The splitting tensile strength of normal-strength SFRC increases with coarse aggregate size up to a maximum of 30 mm and then decreases [9]. For high-strength SFRC, the splitting tensile strength increases with coarse aggregate size up to a maximum of 20 mm and then decreases [9]. With regard to the effect of the length of the steel fibres, it was found that the splitting tensile strength gradually increases as the ratio of fibre length to coarse aggregate size increases up to a maximum of 3, and then decreases [9].

Based on the findings of these studies, it is recommended that the maximum aggregate size for normal-strength SFRC (up to 50 MPa) should be 30 mm, whereas for high-strength SFRC (above 50 MPa) the maximum aggregate size should be 20 mm to achieve the maximum compressive strength. In addition, the optimum split tensile strength is achieved when fibres with a length of three times the aggregate size are used. With regard to the workability, the slump of fresh SFRC increases as the fibre length increases from 30 mm to 60 mm and as the aggregate size increases from 10 mm to 40 mm [9].

1.2.3 Fibre types

The type and the geometrical characteristics of the fibres are key factors for the performance of FRC and the effect of these parameters is developed in the following sections.

The main types of fibres currently utilised in the FRC used in construction are steel, glass and polyvinyl alcohol (PVA) fibres (Figure 1.3), although there are various other types such as polypropylene, polyester macro and micro synthetic, natural fibres and cellulose fibres.

Figure 1.3 Steel, glass and polyvinyl alcohol (PVA) fibres

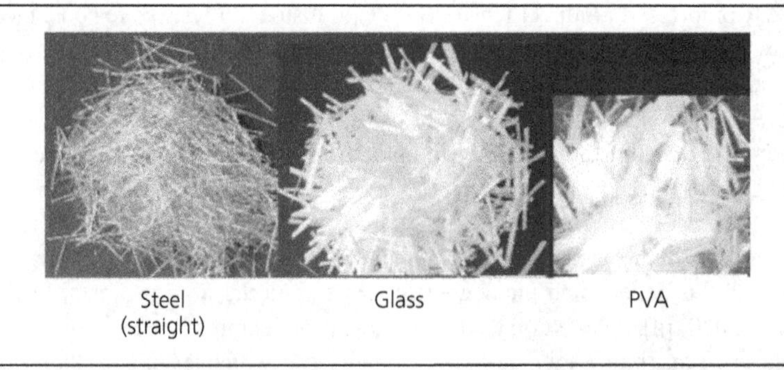

Steel fibres (Figure 1.3) are the most popular type used in construction as they can provide significant enhancement of concrete flexural and tensile strength, toughness and impact resistance and, at the same time, concrete shrinkage and potential cracking can also be addressed.

PVA and glass fibres (Figure 1.3) work in a similar way to the steel fibres but, in general, they are not as effective for enhancement of the flexural and tensile strength characteristics.

Comparative studies [10] have shown that the addition of 1% (per volume) of steel fibres in normal-strength concrete increased the flexural strength by 51% while the respective increment for the same mix using the same volume of glass and polypropylene fibres was found to be 32% and 23%, respectively. This difference was attributed to the enhanced bond of the 'hooked-end' fibres with the cementitious matrix [10].

Fibres have different shapes and lengths. The shape of the fibres is one of the key characteristics for the behaviour of FRC.

There are various geometries of fibres such as the ones presented in Figure 1.4, which, from left to right, shows straight, triple-hooked (5D), double-hooked (4D), hooked-end (3D), corrugated and paddled fibres. The straight fibres rely entirely on the bond between the concrete matrix and the surface of the steel fibres and, therefore, they are prone to slip; pull-out of the fibres is a typical failure mode observed across the cracks. The hooked-end fibres are more effective after the initiation of slip and, therefore, they are more effective in the post-cracking region of SFRC. On the other hand, there are various issues during the mixing process, especially in cases when a large number of fibres with a high aspect ratio (l/d) is used (where l is the length and d is the diameter).

Experimental studies were conducted using various types of fibres with different aspect ratios (40, 60, 80) and dosages (0%, 0.5%, 1.0%, 1.5%, 2.0%) [11] and the results show that as the aspect ratio and the fibre volume are increased, flexural performance is enhanced [11].

Figure 1.4 Various fibre geometries

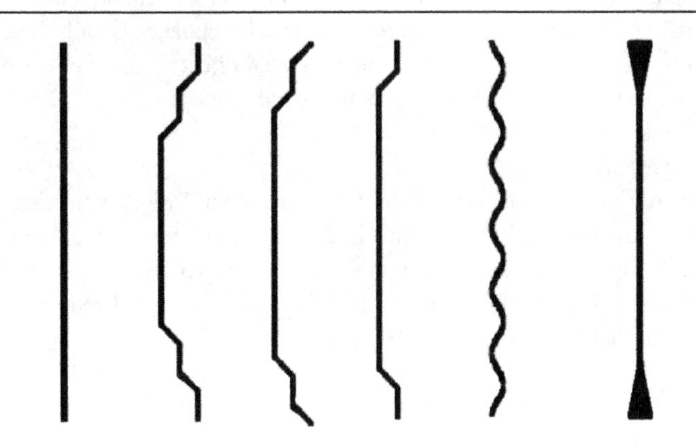

Figure 1.5 Fibre balling [13]

Special attention should also be given to avoid fibre balling (Figure 1.5), particularly when high aspect ratios and high fibre volume fractions are used. More specifically, it was found that the risk of fibre balling in the concrete mixture is reduced when fibres with aspect ratios less than 100 are used [12]. In addition, it was found that the risk of fibre balling is increased when the fibre volume fraction exceeds 1% [12].

When the shape of the fibres is considered, the hooked-end fibres show higher flexural strength and enhanced post-cracking behaviour when compared with straight and corrugated fibres. The hooked-end ones were proved to be the most efficient for enhancement of the flexural strength and the post-cracking performance. Comparisons between hooked-end, double-hooked and triple-hooked fibres using volume fractions of 0.5% and 1% have also been made [13], and the results show that, as the number of hooks was increased, both the ultimate tensile load and the post-cracking performance were substantially enhanced, with the triple-hooked being found to be the most efficient type. The beneficial effect of the number of hooks was more pronounced as the number of fibres was increased.

1.2.4 Mix design process

The mix design process, which is normally followed in case of SFRC, is the same as that used for concrete without fibres and the mixture is designed for specific requirements of compressive strength. Adjustments may be required when a high volume fraction of fibres is used (normally more than 1–2%) to improve the workability. Also, the use of coarse aggregates with large diameters may have a detrimental effect [14].

The range of proportions for normal-weight SFRC for different values of maximum aggregate size are proposed in Table 1.1 [8].

Table 1.1 Typical values for the mix design of SFRC [8]

	Mix design (kg/m³)		
Cement (kg/m³)	400	440	470
Coarse aggregates: 10–15 mm	338		
Coarse aggregates: 10–12 mm	–	155	–
Coarse aggregates: 4–10 mm	534	584	719
Fine aggregates: 0–4mm	909	970	953
Steel fibres (vol. %)	1–2%	1–2%	1–2%
Water over cement ratio	0.55	0.55	0.55

Adjustments may be required for the enhancement of the workability, in particular when self-compacted concrete (SCC) is required. Experimental and numerical work was conducted to evaluate the effect of the presence of fibres on the characteristics of SCC and a design methodology has also been proposed [15–19].

The mix design process has been summarised in the literature [15] and is presented below.

Using a target cubic compressive (f_{cu}) and tensile strength (f_{ft}), the steel fibre volume fraction (v_f) can be calculated using Equations 1.1 and 1.2.

$$f_{ft} = (1 + a_{tb} \times a_{te} \times \lambda_f) f_t \tag{1.1}$$

$$f_t = \frac{(0.65 \times f_{cu} - 8)^{2/3}}{3} \tag{1.2}$$

where:

f_t is the tensile strength of a SCC without fibres
α_{te} is a coefficient linked to fibre distribution
α_{tb} is a coefficient linked to the effect of fibre bridging on the tensile strength
λ_f is the fibre factor which is defined as follows: $v_f \times \frac{l_f}{d_f}$, where l_f is the length and d_f is the diameter of the fibres.

The water to binder ratio can be calculated using Equation 1.3.

$$w/b = \frac{\alpha_a \times f_b}{f_{cu,0} + \alpha_a \times \alpha_b \times f_b} \tag{1.3}$$

where:

α_a and α_b are empirical coefficients linked to the aggregate and concrete type (recommended values: 0.53 and 0.2 for crushed stone [20])
f_b is the compressive strength of the binder material at 28 days.

Table 1.2 Range of factors to consider the effect of fly ash and slag (indicative values using published data [20])

Fly ash content (%)	γ_f
0–40	1.00–0.55
Slag content (%)	γ_s
0–50	1.00–0.70

Table 1.3 Water content range (indicative values using published data [20])

Slump (mm)	**Indicative water content range (kg/m^3)**
10–90	$185 \times c_a - 215 \times c_a$
Aggregate size	c_a
20 mm	1.00
40 mm	0.90

The compressive strength of the binder (f_b) at 28 days can be estimated using Equation 1.4 [20].

$$f_b = \gamma_f \times \gamma_s \times f_{ce} \tag{1.4}$$

where:

γ_f and γ_s are factors for the effect of fly ash and slag, respectively (Table 1.2)
f_{ce} is the strength of the binder.

The strength of the binder can be calculated using Equation 1.5 [20].

$$f_{ce} = \gamma_c \times f_{ce,g} \tag{1.5}$$

where:

γ_c is a coefficient which depends on the type of cement and is taken as 1.12, 1.16 and 1.10 for cements 32.5, 42.5 and 52.5, respectively
$f_{ce,g}$ is the strength grade of the cement.

The selection of water is dependent on the requirements for the specified slump and on the aggregate size and type. Appropriate values can be determined using the information in Table 1.3.

Table 1.4 Recommended range of values for sand percentage (indicative values using published data [20])

Water over cement ratio	Mass of sand/mass of total aggregates (fine and coarse) (%) (indicative range of values)
0.40–0.70	$32 \times c_a - 41 \times c_a$
Aggregate size	c_a
20 mm	1.00
40 mm	0.94

The binder amount, which consists of the total amount of aggregates and cement, is calculated using the w/b ratio, which is given in Equation 1.3.

The sand ratio can be calculated using the values given in Table 1.4.

For steel fibre SCC, a different procedure is proposed for the calculation of the binder.

The mass of the binder can be calculated using Equation 1.6 [15].

$$m_{bf} = (1 + \alpha_{VC} \times \lambda_f) \times m_{b0} \tag{1.6}$$

where:

m_{bf} is the binder content of steel fibre SCC
m_{b0} is the binder content of the unreinforced SCC
α_{VC} is a parameter affected by aggregate size and fibre type with recommended values 0.27 and 0.33 for hooked-end and crimped fibres, respectively [15].

1.3. Mechanical performance and durability

1.3.1 Compressive behaviour

The compressive strength characteristics are not significantly affected by the presence of fibres. Small increments of the compressive strength values in the range of 5–20% can be observed for volume fractions up to 4% [21–23]. In this range of volume fraction, the compressive strength normally increases as the fibre amount is increased. The aspect ratio of the fibres also affects the compressive strength characteristics and it was found that the compressive strength is increased when the aspect ratio is increased from 45 to 65. In addition, for a further increment of the aspect ratio from 65 to 80, the compressive strength is increased for volume fractions up to 1%, whereas, for higher volume fractions, a drop in the compressive strength was observed [22].

Similar observations have also been found for the use of polypropylene fibres where an even higher compressive strength increment of around 28% was obtained for a fibre volume fraction of 1.5%, which was found to be the optimum value. For higher values of the volume fraction, more specifically for 2%, a drop in the compressive strength value was observed. When compared with the same amount of steel fibres, polypropylene fibres were found to be more efficient [24].

The compressive strength of SFRC can be estimated using Equation 1.7 [25].

$$f_{fcu} = f_{cu} \times (1 + \alpha_f \times \lambda_f) \tag{1.7}$$

where:

α_f is taken to be equal to 0.21 for straight, 0.27 for hooked-end and 0.4 for corrugated fibres, respectively [25].

1.3.2 Flexural and tensile behaviour

The flexural and tensile behaviour of FRC is normally significantly enhanced compared with the respective characteristics of plain concrete. The ultimate flexural strength is normally increased by up to 40% when the fibre volume fraction is in the range 0.5–2% [24]. The use of polypropylene fibres and of steel fibres have been examined, and it was found that the polypropylene fibres can be more effective for volume fraction up to 1.5%, whereas 2% steel fibres were found to work better [24].

The geometry of the fibres is also a key parameter for the flexural performance of SFRC. A corrugated profile was shown to be quite effective for the enhancement of flexural strength owing to the enhanced interface conditions, whereas straight steel fibres were found to be the least effective. However, in all the examined cases, the flexural strength increases as the volume fraction is increased [25, 26].

With regard to the toughness and the residual strength, hooked-end fibres were found to be the most efficient type [26].

For the evaluation of the tensile stress–strain characteristics and the effectiveness of the use of fibres, standard flexural tests (Figure 1.6) are recommended in addition to various approaches for the determination of characteristic points of the stress–strain behaviour [27–30]. One of the main characteristics that is evaluated from these tests is the post-cracking behaviour, since the flexural strength can be quite high even for large values of the crack opening. A typical failure mode of FRC prisms at the end of a flexural test is illustrated in Figure 1.7.

In the absence of flexural tests, there is a simplified approach for estimating the flexural strength from the compressive strength values using Equations 1.8–1.11 [27].

Figure 1.6 Indicative flexural test set-up

Figure 1.7 Indicative failure mode of FRC prisms at the end of a flexural test

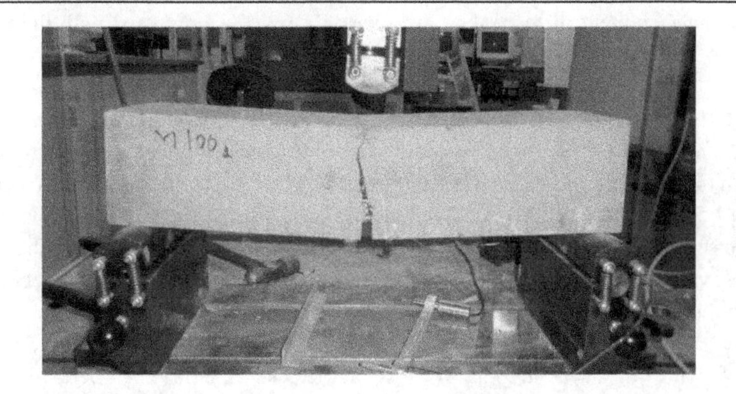

$$f_{\mathrm{fctm}} = 0.3 \times (f_{\mathrm{ck}})^{2/3} \tag{1.8}$$

$$f_{\mathrm{fctk}} = 0.7 \times f_{\mathrm{fctm}} \tag{1.9}$$

$$f_{\mathrm{fctk,fl}} = 1.67 \times f_{\mathrm{fctk}} \tag{1.10}$$

$$f_{\mathrm{fctm,fl}} = 1.43 \times f_{\mathrm{fctk,fl}} \tag{1.11}$$

where f_{fctm} and f_{fctk} are the mean and characteristic value of the tensile strength, respectively, and $f_{\mathrm{fctm,fl}}$ and $f_{\mathrm{fctk,fl}}$ are the respective values of the flexural strength.

1.3.3 Shear behaviour

The shear strength of FRC can be significantly enhanced by the presence of the fibres. Tests were conducted either using a direct shear test set-up [31–33] or by testing beams under shear [34–40].

The results of the direct shear test show that the presence of steel fibres at a weight of more than 40 kg/m^3 (volume 0.5%) leads to quite significant enhancement of shear strength, which increases as the fibre volume fraction is increased [31]. Another investigation on the use of PVA fibres shows that the shear strength is optimum for 0.1% fibre contact and is reduced for higher fibre contents. This is attributed to the effect of the fibres on the binder-fibre bond [32].

Double-notch and push-off tests have also been conducted to evaluate the effect of the steel fibre volume and the effect of the size of the aggregates on the shear strength of SFRC (Figure 1.8) [33]. In case of push-off tests, transverse reinforcement was also used in addition to the steel fibres.

The results show that for small values of the volume fraction (0.5%) the shear strength is not significantly affected by the presence of steel fibres, but when a higher volume of fibres was used (1.5%) the shear strength was significantly increased and it was found to be more than double that of the respective specimens without steel fibres. The effect of the coarse aggregates was also found to be quite important, and when aggregates with a maximum size of 12.5 mm were used, an increment of more than 50% was observed as compared with the respective results of specimens with a maximum aggregate size of 9.5 mm.

The results of push-off tests, where the presence of transverse reinforcement was also examined, show that the specimens without steel fibres reached the shear strength peak with yielding of transverse reinforcement in contrast to the specimens with steel fibres (with either a 0.5% or a 1.5% volume fraction) for which the peak shear stress was reached without yielding of transverse reinforcement. The specimens with 1.5% of steel fibres show enhanced peak shear stress values that are significantly higher than the respective results of the specimens with 0.5% steel fibres or without steel fibres [33].

The calculation of the shear strength of SFRC at a cracked interface can be expressed using Equations 1.12 and 1.13 [33, 41].

$$\tau_{\mathrm{SFRC}} = \tau_{\mathrm{c}} + \mu \times \sigma_{\mathrm{N}} + \tau_{\mathrm{Fibres}} \tag{1.12}$$

where:

τ_{c} is the shear strength of cracked plain concrete
μ is the friction coefficient of the cracked shear plane
σ_{N} is the normal stress to the interface
τ_{Fibres} is the contribution of steel fibres crossing the cracked plane.

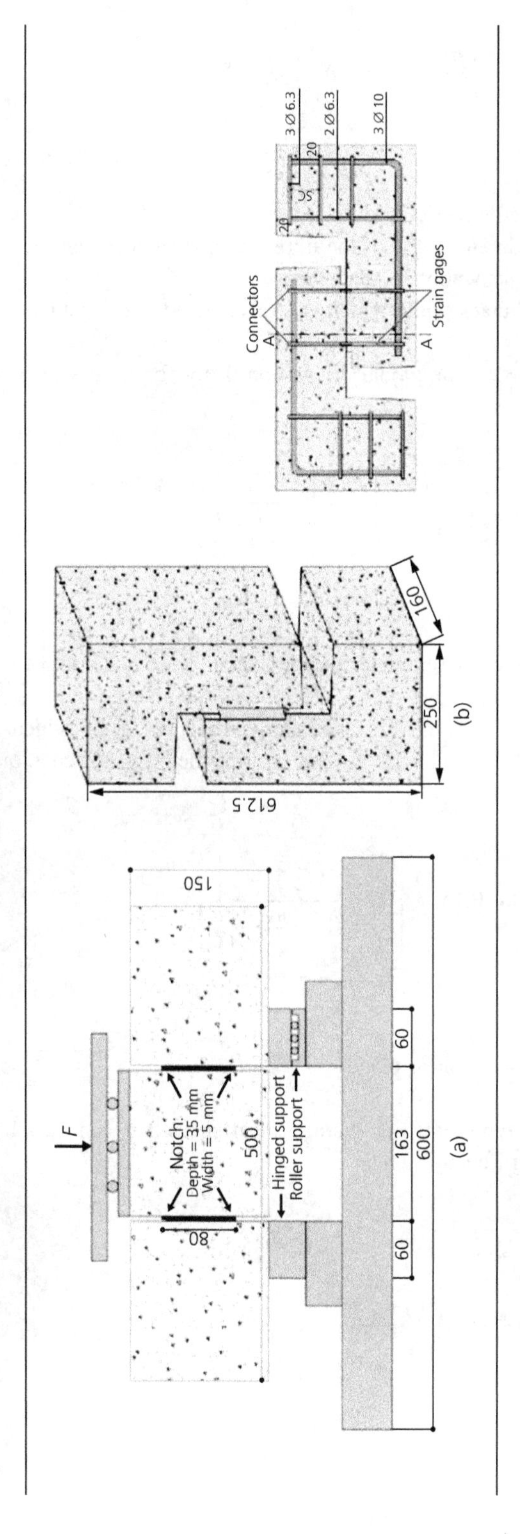

Figure 1.8 (a) Double-notch and (b) push-off tests used to evaluate the shear strength of SFRC [33]

$$\tau_{\text{Fibres}} = \frac{2 \times K_\tau \times \rho_f \times K_b}{d_f} \times \tau(\delta) \times (l_a - \delta - w_o) \tag{1.13}$$

where:

ρ_f is the fibre volumetric ratio
l_a is the embedment length of the fibre which is taken as half of the fibre length
δ is the crack separation displacement
w_o is the initial crack width which it is taken as 0 for an uncracked shear plane.

The coefficient K_τ is the global orientation factor for shear which is calculated using Equation 1.14.

$$K_\tau = \frac{2}{\pi} \times \tan^{-1}\left[\sqrt{\frac{\delta + w_o}{\alpha_{II} \times l_f}}\right] \times \left(1 - \frac{2 \times s}{l_f}\right) \tag{1.14}$$

where:

α_{II} is the engagement parameter which is taken to be equal to $3\frac{3 \times d_f}{2 \times l_f}$

The coefficient K_b is the fibre-boundary influence factor which considers the boundary conditions applied to the fibres that restrict their free orientation. It is calculated using Equation 1.15.

$$K_b = \frac{4}{\pi} \times \left\{1 - 0.15 \times \left[0.5 + \frac{\frac{h}{l_f} - 1}{1 + \left|\frac{h}{l_f} - 1\right|}\right]\right\} \geq 1 \tag{1.15}$$

where:

h is the width of the shear plane.

The factor $\tau(\delta)$ represents the average shear stress between the fibre and the matrix and is calculated using Equation 1.16.

$$\tau_\delta = \begin{cases} \tau_{max}\left(\frac{\delta}{s_1}\right)^a & \text{if } \delta \leq s_1 \\ \tau_{max} - \frac{(\tau_{max} - \tau_f)(\delta - s_1)}{s_2 - s_1} & \text{if } s_1 \leq \delta \leq s_2 \\ \tau_f & \text{if } \delta > s_2 \end{cases} \tag{1.16}$$

Proposed parameters for Equation 1.16 are given in the literature [42] and the recommended values are: $\tau_{max} = 6$ MPa, $\alpha = 0.4$, $\tau_f = 0.6$ MPa and $s_1 = 0.01$ mm. For s_2, a value of 6.5 mm is recommended for fibre volume fractions of 1.5% and above and 1 mm for fibre volume fractions of 0.5% and below. For fibre volume fractions between 0.5% and 1.5%, linear interpolation is suggested [42].

The effectiveness of the use of fibres for the shear enhancement of RC beams has also been extensively studied [34–40]. The results show that the presence of fibres can lead to shear strength enhancement, which can also lead to a change of the failure mode of RC beams from shear to flexural failure. Enhancement of the ductility of the beams was also achieved, which increases as the fibre volume fraction is increased.

It has also been confirmed that the presence of fibres has the potential to be used as partial replacement of the stirrups of RC beams [35].

Experimental results from one of these studies [39] show that the load-carrying capacity of the beam was increased by 12% when shear links were replaced by hooked-steel fibres, and when crimped steel fibres were used, this increment was found to be 15% [39].

From these results, it was concluded that 1% of crimped steel fibres of aspect ratio (50) was more efficient for the replacement of shear links compared with hooked-end steel fibres [39].

An equation was proposed by Concrete Society TR63 [28] for the calculation of the shear resistance (V_{rd}) of cross-sections with shear links and fibres (Equation 1.17).

$$V_{rd} = \left[\left(\frac{0.18}{\gamma_c}\right) \times k \times (100 \times \rho_1 \times f_{ck})^{\frac{1}{3}} + v_{fd}\right] \times b \times d + V_{wd} \quad (1.17)$$

where:

γ_c is the safety factor which is taken as 1.5 for concrete
$k = 1 + (200/d)^{1/2} \leq 2$ (where d is the effective depth of the cross-section)
$\rho_1 = \frac{A_{s1}}{b \times d} \leq 0.02$ (where A_{s1} is the cross-sectional area of the tensile reinforcement)
b is the cross-sectional width
V_{wd} is the shear links contribution
v_{fd} is the shear strength contribution of fibres.

V_{wd} and v_{fd} can be calculated using Equation 1.18 (BS EN 1992-1-1 [43]) and Equation 1.19, respectively.

$$V_{wd} = 0.9 \times d \times \left(\frac{A_{sw}}{S}\right) \times f_{ywd} \quad (1.18)$$

where:

A_{sw} is the cross-sectional area of the links
S is the links spacing
f_{ywd} is the yield strength of the links.

$$v_{fd} = 0.7 \times k_f \tau_{Fibres} \tag{1.19}$$

where:

k_f is a factor that considers the effect of flanges in flanged cross-sections and is taken as 1.0 for rectangular ones.

1.3.4 Punching shear of SFRC slabs

The contribution of fibres is important for the punching shear performance of SFRC slabs.

There are semi-empirical models available that can be used for the estimation of the punching shear resistance of SFRC slabs. One of the main models used to estimate the punching shear capacity of SFRC slabs is that proposed by Narayanan and Darwish (Equation 1.20) [44].

$$\frac{V_u}{b_{pf} \times d} = \lambda_s \times \left(0.24 \times f_{spf} + 16 \times \rho + v_b\right) \tag{1.20}$$

where:

b_{pf} is the modified perimeter of the critical section in case of FRC slabs and is calculated using the following equation.

$$b_{pf} = b_{0,1.5d} \times (1 - 0.55 \times F)$$

where:

$b_{0,1.5d}$ is located at a distance of 1.5 times the column side from the face of the column.

d is the effective depth
ρ is the longitudinal reinforcement ratio
f_{spf} is the maximum split cylinder strength of the SFRC and can be calculated using the following equation.

$$f_{spf} = \frac{f_{ck}}{20 - \sqrt{F}} + 0.7 + \sqrt{F}$$

F is the fibre parameter and can be calculated using the following equation.

$$F = \frac{k_b \times l_f \times \rho_f}{d_f}$$

where:

k_b is the bond factor
l_f and d_f are the length and the diameter of the fibres, respectively
ρ_f is the fibre volume fraction

v_b is the vertical fibre pull-out stress along the inclined crack and can be calculated using the following equation.

$$v_b = 0.41 \times \tau \times F$$

where:

τ is the average fibre–matrix interfacial bond stress with proposed value 4.15 N/mm^2

λ_s is the size effect factor and is calculated using the following equation.

$$\lambda_s = 1.6 - 0.002 \times h$$

where h is the height of the slab.

A more advanced model was later proposed by Higashiyama *et al.* [45] which takes into consideration the contribution of the fibres based on their pull-out strength. The basic control perimeter is calculated depending on the amount and the fibre properties (Equation 1.21).

$$\frac{V_u}{b_{pf} \times d} = \beta_d \times \beta_p \times \beta_r \times (f_{pcd} + v_b) \tag{1.21}$$

where:

$$b_{pf} = (u + \pi \times d) \times (1 - 0.32 \times F)$$

and

u is the perimeter of the loading area.

$$f_{pcd} = 0.2 \times \sqrt{f_c} \leq 1.2 \text{ MPa}$$

$$\beta_d = \sqrt[4]{\frac{1000}{d}} \leq 1.5$$

$$\beta_p = \sqrt[3]{100\rho} \leq 1.5$$

A mechanical model has also been presented by Maya *et al.* [46], which can be used to calculate the punching shear resistance of slabs with conventional steel bars in addition to steel fibres.

The use of a mechanical model of the critical shear crack theory is proposed for the calculation of the ultimate punching shear strength and deformation capacity of RC slabs.

According to this theory, the strength of the diagonal concrete compressive strut that is carrying the shear actions is reduced as the shear crack opening is increased, where the shear crack opening is considered to be proportional to the rotation of the slab (ψ).

Based on this theory, the punching shear resistance of the slabs with steel fibre reinforcement (V_R) can be calculated using Equation 1.22.

$$V_R = V_{R,c} + V_{R,f} \tag{1.22}$$

where:

$V_{R,c}$ is the punching shear resistance of slabs without transverse reinforcement and without steel fibres
$V_{R,f}$ expresses the contribution of steel fibres to the shear strength.

Equation 1.23 was proposed for the calculation of the punching shear resistance of slabs without transverse reinforcement and without steel fibres ($V_{R,c}$).

$$\frac{V_{R,c}}{b_o \times d \times \sqrt{f_c}} = \frac{2}{3 \times \gamma_c} \times \frac{1}{1 + 20 \times \frac{\psi \times d}{d_{g0} + d_g}} \tag{1.23}$$

where:

Ψ is the slab rotation
γ_c is concrete safety factor equal to 1.5
b_o is the control perimeter at a distance of $d/2$ from the face of the column
f_c is the compressive strength of concrete
d_g is the aggregate size
d_{g0} is a reference aggregate size equal to 16 mm.

The crack opening and the respective fibre bridging stresses at the critical failure area are taken into consideration when calculating the contribution of the steel fibres ($V_{R,f}$) (Figure 1.9).

Integration of the fibre bridging stresses of each crack width across the critical failure (which is considered to be formed at an angle of 45° from the soffit of the slab) can be used to calculate the contribution of the steel fibres ($V_{R,f}$) (Equation 1.24).

$$V_{R,f} = \int_{A_p} \sigma_{tf} \times (w(\xi))dA_p \tag{1.24}$$

Figure 1.9 Crack width and fibre bridging stress development along the shear crack of the slabs [46]

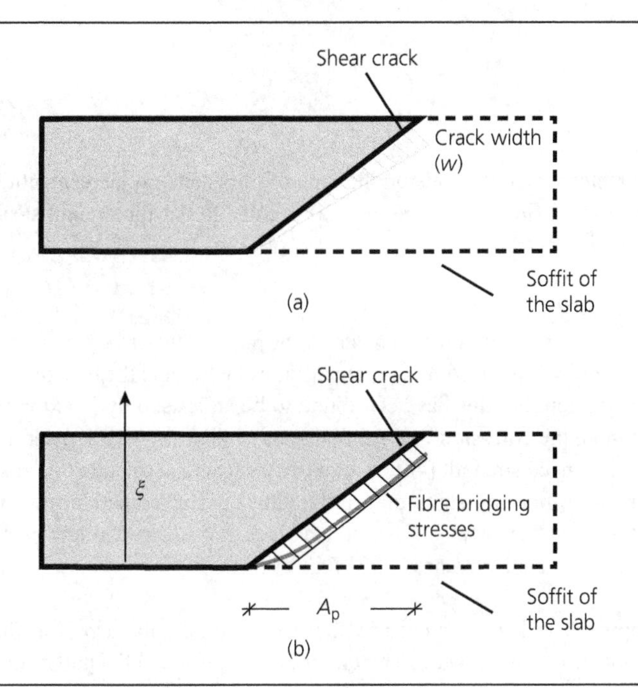

where:

σ_{tf} is the fibre bridging stresses
w is the crack width
A_p is the horizontally projected area of the punching shear failure surface (Figure 1.9)
ξ is the vertical distance from the sofit (Figure 1.9).

According to the proposal of Muttoni and Fernández Ruiz [47], the average bridging stress can be estimated from the crack opening at a distance $d/3$ from the soffit of the slab and can be calculated using Equation 1.25.

$$V_{R,f} = A_p \times \sigma_{tf}\left(w = \frac{\psi \times d}{6}\right) \tag{1.25}$$

where:

$$\sigma_{tf} = K_f \times \alpha_f \times \rho_f \times \tau_b$$

$$K_f = \frac{1}{\pi}\arctan\left(\alpha_e \times \frac{w}{d_f}\right) \times \left(1 - \frac{2 \times w}{l_f}\right)^2$$

α_e is an engagement parameter with proposed value 3.5
τ_b is the bond strength between the cementitious matrix and the fibres.

The proposed equation for the bond strength (τ_b) is

$$\tau_b = k_b \times \sqrt{f_c}$$

where:

k_b is the bond factor which depends on the type of fibres and on the cementitious matrix. The proposed value (in the absence of data) is equal to 0.4 for straight steel fibres and 0.6 for crimped steel fibres.

1.3.5 Durability

Concrete durability is an important parameter for the performance of concrete structures. In the case of FRC, durability is affected by the cementitious matrix and the type of fibre. Steel fibres are subject to corrosion, but this has been found to be an issue only in extremely aggressive environments where the corrosion of steel fibres may initiate concrete cracking and a subsequent reduction in tensile strength [48]. In other cases where accelerated corrosion tests were conducted there were not any signs of deterioration or cracks apart from some superficial rusting, which shows that steel reinforcement does not corrode under wet–dry chloride exposure and proves the effectiveness of SFRC [49].

A common approach to better understand the deterioration approach is to differentiate the initiation from the propagation phase. The initiation stage is the initial part of the process from the depassivation of the steel due to the pH reduction and the ingress of the chloride ions through the cementitious matrix or the cracks [50]. Then the progress of this process, and potentially the localisation to specific parts, is the propagation stage.

A conceptual model was presented in the literature [50] which shows the deterioration due to corrosion of cracked and uncracked SFRC compared with uncracked RC [50].

Figure 1.10 shows that uncracked concrete behaves in the same way for RC and SFRC at the initiation stage, whereas, at the propagation stage, the rate of the damage increment is much higher for RC which shows the beneficial effect of the fibres for the control of damage.

In the case of SFRC, the damage is mostly linked to the reduction of the cross-sectional area of the steel fibres and normally there is not corrosion-induced cracking or spalling in contrast to RC.

In the case of cracked SFRC, the behaviour is similar to that of uncracked SFRC, with the main difference being that higher damage is expected.

The performance of FRC with various types of fibres has been studied under various deterioration processes such as chlorides, carbonation, alkali–silica reaction, high temperatures, and freeze and thaw tests, and the main conclusion is that, in general, durability is enhanced in the case of FRC owing to the crack-control feature of the fibres which limits the deterioration rate and the entry of substances such as water, chlorides and carbon dioxide [51].

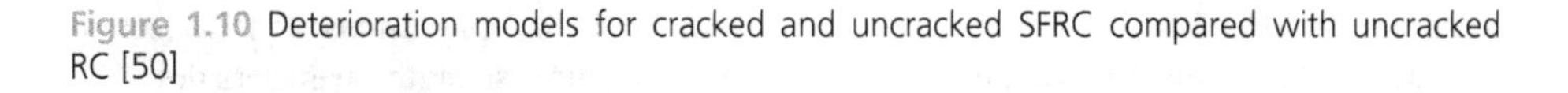

Figure 1.10 Deterioration models for cracked and uncracked SFRC compared with uncracked RC [50]

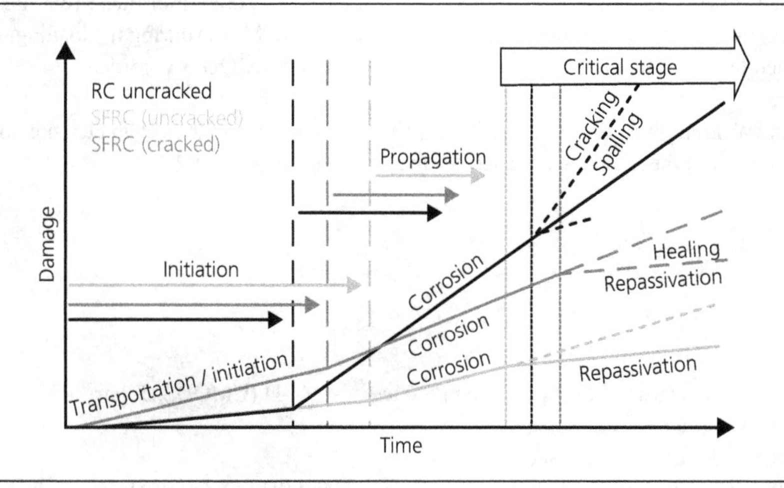

The effect of acid attack on SFRC has also been studied [52] and the results have shown that the damage to uncracked SFRC exposed to acids is similar to that for plain concrete as the corrosion damage of steel fibres is limited to the fibres enclosed within the chemically eroded concrete [52]. When there is severe exposure of cracked and uncracked SFRC to acids, significantly higher deterioration is observed compared with other types of exposure such as carbonation or chloride exposure. However, this may be attributed to the deterioration of the surrounding concrete rather than the corrosion of the steel fibres and the need for further research in this direction was highlighted [52].

1.4. Design of structural elements using FRC

The general approach that is used for the design of FRC elements is based on the stress–strain method [27, 30], which involves the same principles as used in the design of conventional RC elements. The main difference in the design of the FRC elements is linked to the enhanced tensile stress–strain characteristics that are taken into consideration, in contrast to the conventional RC elements for which the contribution of concrete in tension is neglected.

1.4.1 Tests to determine the FRC characteristics

The recommended procedure for the determination of flexural and tensile characteristics of SFRC is the use of prisms (Figure 1.6 and Figure 1.7) with various geometries, as proposed in relevant code recommendations such as the British Standards (BS EN 14651:2005), Japanese Standards (JCI-SF4), European Standards (EFNARC), German Standards (DIN) and American Standards (ASTME) [28]. The main differences between the types of recommendations are based on the geometry of the examined prisms and, in this document, the British Standards (BS EN 14651:2005) method will be presented [29].

According to BS EN 14651 [29], standard prisms with the dimensions and the set-up of Figure 1.11 should be examined to determine the flexural strength characteristics. The specimens have a 25 mm notch in the middle of the span where a transducer (clip gauge) is placed to measure the crack mouth opening displacement (CMOD) during the loading. A load is applied to the middle of the span and the load with the CMOD is recorded.

The flexural strength is determined using Equation 1.26 for the various characteristic points of Figure 1.12, and these values are used for the characterisation of FRC.

$$f_{\mathrm{fctfl,j}} = \frac{3 \times F_{\mathrm{j}} \times l}{2 \times b \times h_{\mathrm{sp}}^2} \tag{1.26}$$

where:

F_{j} is the load corresponding to different values of CMOD ($\mathrm{CMOD_j}$) (Figure 1.12)
l is the span length (i.e. 500 mm)
b is the width of the prism (i.e. 150 mm)
h_{sp} is the distance between the top of the specimen and the beginning of the notch (i.e. 125 mm).

Figure 1.11 Experimental set-up for the flexural tests of SFRC prisms [29, 30]. (Reproduced from *fib Model Code for Concrete Structures* (2010) Chapter 5 Materials, page 145, Figure 5.6-5: Test set-up required by EN 14651 (dimensions in [mm]), with permission from the International Federation for Structural Concrete (*fib*)). Permission to reproduce extracts from British Standards is granted by BSI Standards Limited (BSI). No other use of this material is permitted. British Standards can be obtained from BSI Knowledge knowledge.bsigroup.com

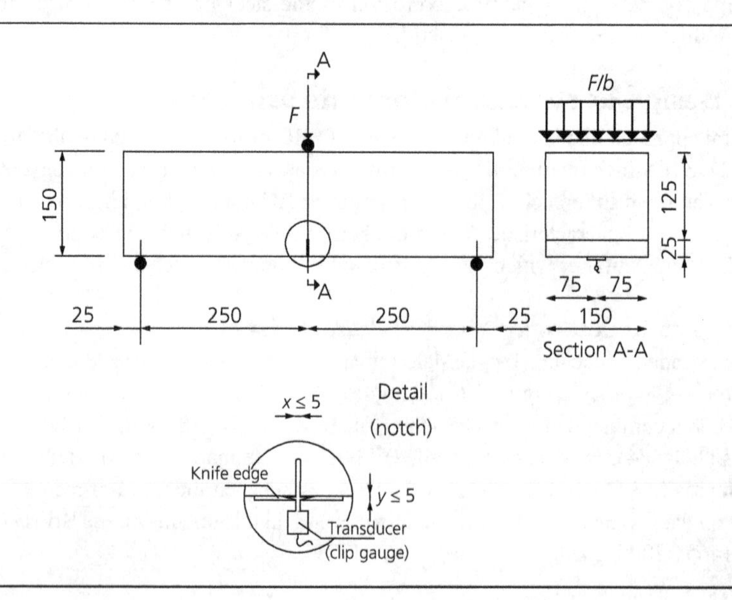

Figure 1.12 Load–CMOD graph for SFRC flexural tests [29, 30]. (Reproduced from *fib Model Code for Concrete Structures* (2010) Chapter 5 Materials, page 146, Figure 5.6-6: Typical load *F*-cmod curve for plain concrete and FRC, with permission from the International Federation for Structural Concrete (*fib*)). Permission to reproduce extracts from British Standards is granted by BSI Standards Limited (BSI). No other use of this material is permitted. British Standards can be obtained from BSI Knowledge knowledge.bsigroup.com

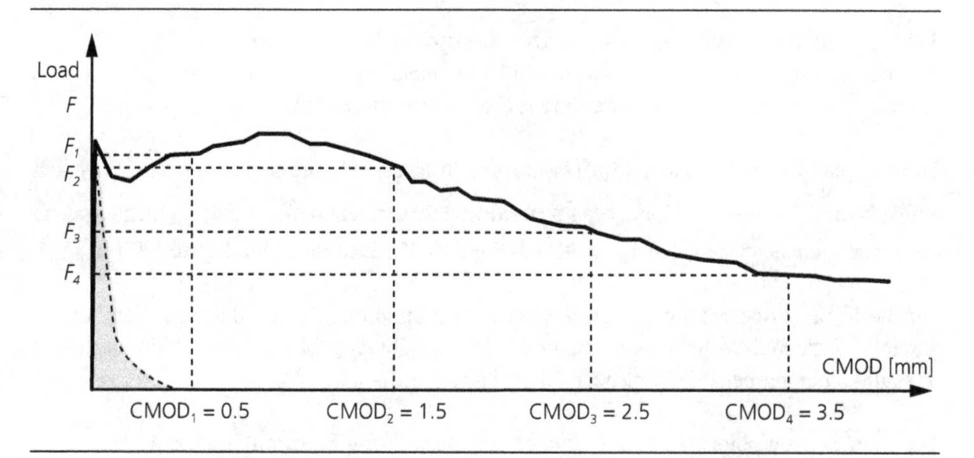

1.4.2 FRC classifications

The flexural strength at the different CMOD values provides useful information about the performance of SFRC and can also be used for its classification [30].

According to the fib Model Code 2010 [30], the classification of SFRC uses a number and a letter that denote the flexural strength $f_{\text{fctl},1}$ (which is the value at $\text{CMOD}_1 = 0.5$ mm) and the residual strength ratio $\frac{f_{\text{ctfl},3}}{f_{\text{ctfl},1}}$, where $f_{\text{fctl},3}$ is the residual strength at $\text{CMOD}_3 = 2.5$ mm.

More specifically, the residual strength ratio can be in the following intervals and is classified using the letters a–e, as specified below (Equation 1.27).

$$
\begin{aligned}
&\text{a for } 0.5 < \frac{f_{\text{ctfl},3}}{f_{\text{ctfl},1}} < 0.7 \\
&\text{b for } 0.7 \leq \frac{f_{\text{ctfl},3}}{f_{\text{ctfl},1}} < 0.9 \\
&\text{c for } 0.9 \leq \frac{f_{\text{ctfl},3}}{f_{\text{ctfl},1}} < 1.1 \\
&\text{d for } 1.1 \leq \frac{f_{\text{ctfl},3}}{f_{\text{ctfl},1}} < 1.3 \\
&\text{e for } 1.3 \leq \frac{f_{\text{ctfl},3}}{f_{\text{ctfl},1}}
\end{aligned}
\tag{1.27}
$$

For example, SFRC 3b has flexural strength ($f_{\text{fctfl},1}$) between 3 and 4 MPa and residual strength ratio $0.7 \leq \frac{f_{\text{ctfl},3}}{f_{\text{ctfl},1}} \leq 0.9$.

The flexural strength and the residual strength characteristics are significantly affected by the fibre volume fraction and by the length, aspect ratio and geometry of the fibres.

1.4.3 Constitutive models for the design of FRC elements

Two models are proposed for the design of FRC elements at ultimate limit state (ULS): one is the rigid plastic model and the other one is the linear post-cracking model [30].

The rigid plastic model (Figure 1.13(a)) assumes a constant tensile-stress–crack-opening (w) graph with a value $f_{\text{Ftu}} = \frac{f_{\text{R3}}}{3}$. This model can be used for the calculation of ultimate moment capacity (M_u) at an ultimate crack value w_u = CMOD (Figure 1.12) as illustrated in Figure 1.13(b).

For the linear post-cracking model, the linear distribution of the tensile-crack opening (w) graph is considered, with an ascending or a descending branch for FRC with strain-hardening or strain-softening behaviour, respectively (Figure 14(a)).

The characteristic points f_{Fts} and f_{Ftu} can be calculated using Equations 1.28 and 1.29.

$$f_{\text{Fts}} = 0.45 \times f_{\text{R1}} \quad (1.28)$$

Figure 1.13 (a) Rigid plastic model and (b) calculation of ultimate moment capacity (M_u) [30]. (Reproduced from *fib Model Code for Concrete Structures* (2010) Chapter 5 Materials, page 147, Figure 5.6-7: Simplified post-cracking constitutive laws: stress-crack opening (continuous and dashed lines refer to softening and hardening post-cracking behaviour, respectively) and Figure 5.6-8: Simplified model adopted to compute the ultimate residual tensile strength in uniaxial tension f_{FTU} by means of the residual nominal bending strength f_{R3}, with permission from the International Federation for Structural Concrete (*fib*))

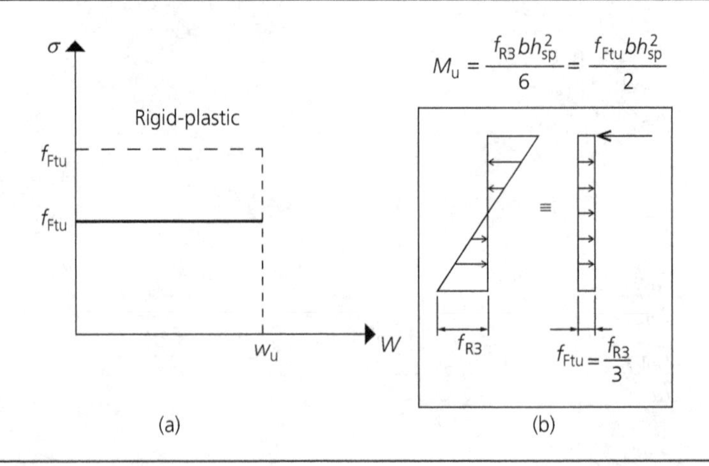

Figure 1.14 (a) Linear post-cracking model and calculation of moment capacity (b) at SLS and (c) at ULS. (Reproduced from *fib Model Code for Concrete Structures* (2010) Chapter 5 Materials, page 147, Figure 5.6-7: Simplified post-cracking constitutive laws: stress-crack opening (continuous and dashed lines refer to softening and hardening post-cracking behaviour, respectively) and Figure 5.6-9: Stress diagrams for the determination of the residual tensile strength f_{FTS} (b) and f_{FTU} (c) for the linear model, respectively, with permission from the International Federation for Structural Concrete (*fib*))

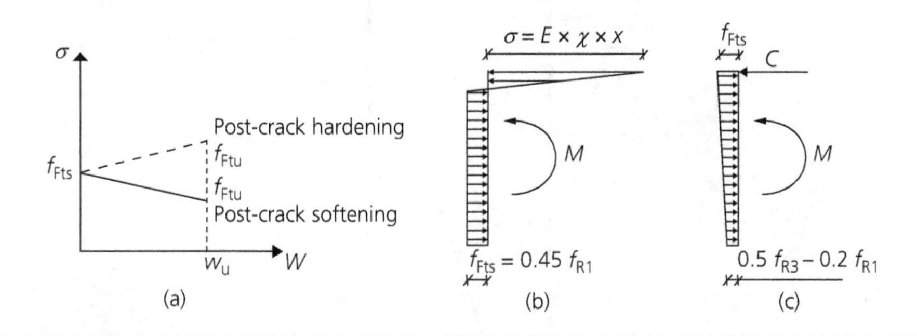

$$f_{Ftu} = f_{Fts} - \frac{w_u}{CMOD_3} \times (f_{Fts} - 0.5 \times f_{R3} + 0.2 \times f_{R1}) \geq 0 \tag{1.29}$$

For the calculation of the moment capacity at serviceability limit state (SLS) ($CMOD_1$) (Figure 1.14(b)) and ULS ($CMOD_3$) (Figure 1.14(c)), rigid linear behaviour is assumed and the moment (M) is calculated using Equation 1.30.

$$M = \frac{f_R \times b \times h_{sp}^2}{6} \tag{1.30}$$

where:

f_R is the flexural strength f_{R3} for $CMOD_3$ (ULS) and f_{R1} for $CMOD_1$ (SLS)

For design purposes, the following safety factors are also proposed for ULS (Table 1.5).

For design at SLS, safety factors should be taken to be equal to 0 [30].

Table 1.5 FRC safety factors for ULS [30]

Condition	FRC safety factor
Tension (residual strength)	1.5
Compression and tension at limit of linearity	Same as plain concrete

Figure 1.15 Strain calculation along the cross-sectional height [30]. (Reproduced from *fib Model Code for Concrete* Structures (2010) Chapter 5 Materials, page 147, Figure 5.6-9: Stress diagrams for the determination of the residual tensile strength f_{FTS} (b) and f_{FTU} (c) for the linear model, respectively, with permission from the International Federation for Structural Concrete (*fib*))

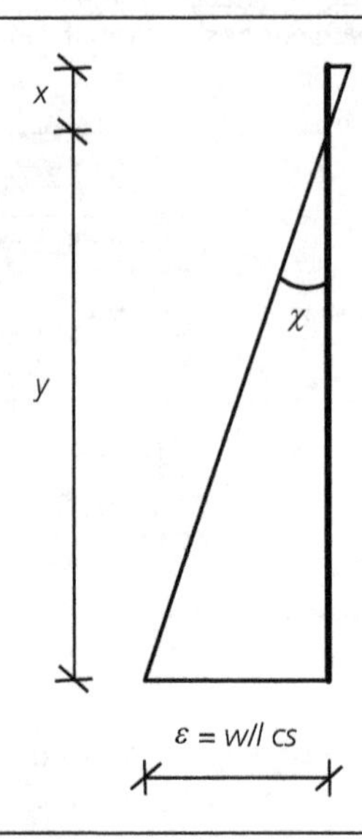

For numerical analyses, the stress–strain constitutive model can be calculated using the linear model of Figure 1.14(a) (Equations 1.28 and 1.29) by converting the crack opening (w) to strain (*TNR*). The strain (ϵ) can be calculated by dividing the crack opening (w) by the structural characteristic length (l_{cs}) (Figure 1.15).

For elements reinforced with steel bars, l_{cs} is calculated using Equation 1.31.

$$l_{cs} = \min\{s_{rm}, y\} \tag{1.31}$$

where:

s_{rm} is the mean distance between cracks
y is the distance between the neutral axis and the tensile side of the cross-section.

REFERENCES

1. Kaur P and Talwar M (2017) Different types of fibres used in FRC. *International Journal of Advanced Research in Computer Science* **8(4)**: 380–383.
2. Naaman A (2018) Fibre reinforced concrete: five decades of progress. *Proceedings of the 4th Brazilian Conference on Composite Materials, Rio de Janeiro*, pp. 1–22.
3. Zollo R (1997) Fiber-reinforced concrete: an overview after 30 years of development. *Cement and Concrete Composites* **19**: 107–122.

4. Simões T, Octávio C, Valença J, Costa H, Dias-da-Costa D and Júlio E (2017) Influence of concrete strength and steel fibre geometry on the fibre/matrix interface. *Composites Part B: Engineering* **122**: 156–164.
5. Seleem MH, Badawy A and Mohamed SS (2020) Effect of maximum aggregate size on the mechanical properties and flexural energy of FRC. *Al-Azhar University Civil Engineering Research Magazine (CERM)* **42(1)**: 86–94.
6. Ulas MA, Alyamac KE and Ulucan ZC (2017) Effects of aggregate grading on the properties of steel fibre reinforced concrete. *Fibre Concrete 2017 IOP Conference Series: Materials Science and Engineering* **246**: 012015.
7. Meddah MS, Zitouni S and Belâabes S (2010) Effect of content and particle size distribution of coarse aggregate on the compressive strength of concrete. *Construction and Building Materials* **24**: 505–512.
8. Olivito RS and Zuccarello FA (2010) An experimental study on the tensile strength of steel fiber reinforced concrete. *Composites Part B: Engineering* **41**: 246–255.
9. Han J, Zhao M, Chen J and La X (2019) Effects of steel fiber length and coarse aggregate maximum size on mechanical properties of steel fiber reinforced concrete. *Construction and Building Materials* **209**: 577–591.
10. Hussain I, Ali B, Akhtar T, Jameel MS and Raza SS (2020) Comparison of mechanical properties of concrete and design thickness of pavement with different types of fiber-reinforcements (steel, glass, and polypropylene). *Case Studies in Construction Materials* **13**: e00429.
11. Li B, Xu L, Shi Y, Chi Y, Liu Q and Changning L (2018) Effects of fiber type, volume fraction and aspect ratio on the flexural and acoustic emission behaviors of steel fiber reinforced concrete. *Construction and Building Materials* **181**: 474–486.
12. Abdallah S, Rees D, Ghaffar S and Fan M (2018) Understanding the effects of hooked-end steel fibre geometry on the uniaxial tensile behaviour of self-compacting concrete. *Construction and Building Materials* **178**: 484–494.
13. Barman M (2018) *Comparison of Performances of Structural Fibers and Development of a Specification for Using Them in Thin Concrete Overlays*. Research project, University of Minnesota, USA.
14. ACI 544.3R-93 (1998) Guide for Specifying, Proportioning, Mixing, Placing, and Finishing Steel Fiber Reinforced Concrete. American Concrete Institute.
15. Ding X, Zhao M, Li J, Shang P and Li C (2020) Mix proportion design of self-compacting SFRC with manufactured sand based on the steel fiber aggregate skeleton packing test. *Materials MDPI* **13**: 2833.
16. Ding X, Li C, Han B, Lu Y and Zhao S (2018) Effects of different deformed steel-fibers on preparation and fundamental properties of self-compacting SFRC. *Construction and Building Materials* **168**: 471–481.
17. Zhao M, Ding X, Li J and Law D (2018) Numerical analysis of mix proportion of self-compacting concrete compared to ordinary concrete. *Key Engineering Materials* **789**: 69–75.
18. Ding X, Zhao M, Zhou S, Fu Y and Li C (2019) Statistical analysis and preliminary study on the mix proportion design of self-compacting SFRC. *Materials MDPI* **12**: 637.
19. Vilanova A, Fernandez-Gomez J and Landsberger G (2011) Evaluation of the mechanical properties of self compacting concrete using current estimating models: estimating the modulus of elasticity, tensile strength and modulus of rupture of self compacting concrete. *Construction and Building Materials* **25**: 3417–3426.
20. Yuan Q, Liu Z, Zheng K and Ma C (2021) *Civil Engineering Materials: From Theory to Practice*. Woodhead Publishing Series in Civil and Structural Engineering, Elsevier.

21. Phanikumar BR and Sofi A (2016) Effect of pond ash and steel fibre on engineering properties of concrete. *Ain Shams Engineering Journal* **7(1)**: 89–99.
22. Yazici S, Inan G and Tabak V (2007) Effect of aspect ratio and volume fraction of steel fiber on the mechanical properties of SFRC. *Construction and Building Materials* **21**: 1250–1253.
23. Mohite DB and Shinde SB (2013) Experimental investigation on effect of different shaped steel fibers on compressive strength of high strength concrete. *IOSR Journal of Mechanical and Civil Engineering* **6(4)**: 24–26.
24. Thiyagarajan R and Pazhani KC (2019) Experimental investigations on the behaviour of PPFRC and SFRC. *International Journal of Advanced Research in Engineering and Technology* **10(3)**: 194–202.
25. Zhang L, Zhao J, Cunyuan F and Wang Z (2020) Effect of surface shape and content of steel fiber on mechanical properties of concrete. *Advances in Civil Engineering* **52(4)**: 70–82.
26. Soulioti DV, Barkoula NM, Paipetis A and Matikas TE (2009) Effect of fibre geometry and volume fraction on the flexural behaviour of steel-fibre reinforced concrete. *Strain* **47(4)**: 535–541.
27. Rilem TC 162-TDF (2003) Test and design methods for steel fiber reinforced concrete. *Recommendations for σ-ϵ-design method. Materials and Structures* **36**: 560–567.
28. The Concrete Society (TCS) (2007) *Guidance for the Design of Steel-Fibre-Reinforced Concrete*. The Concrete Society, Camberley, UK. Technical Report TR63.
29. BS EN 14651 (2005) Test method for metallic fibre concrete — Measuring the flexural tensile strength (limit of proportionality (LOP), residual). BSI, London, UK.
30. fib (2010) Model Code for Concrete Structures 2010. Federation Internationale Du Beton, Ernst & Sohn.
31. Khanlou A, MacRae GA, Scott AN Hicks SJ and Clifton GC (2012) Shear performance of steel fibre-reinforced concrete. *Proceedings of the Australasian Structural Engineering Conference, Perth, Australia*, pp. 1–8.
32. Varghese A, Agrima M, Balakrishnan M, Greeshma C and Krishnapriya PS (2019) Pure shear strength of PVA fiber reinforced concrete. *International Journal of Recent Technology and Engineering* **8(2)**: 217–221.
33. Araújo DL, Lobo FA and Martins BG (2021) A shear stress-slip relationship in steel fibre-reinforced concrete obtained from push-off testing. *Construction and Building Materials* **293**: 123435.
34. Furlan S and Hanai JB (1997) Shear behaviour of fiber reinforced concrete beams. *Cement and Concrete Composites* **19**: 359–366.
35. Abbas AA, Mohsin SMS, Cotsovos DM and Ruiz-Teran AM (2014) Shear behaviour of steel-fibre reinforced concrete simply supported beams. *Proceedings of the Institution of Civil Engineers – Structures and Buildings* **167**: 544–558.
36. Dinh HH, Parra-Montesinos GJ and Wight JK (2010) Shear behavior of steel fiber reinforced concrete beams without stirrup reinforcement. *ACI Structural Journal* **107(5)**: 597–606.
37. Mansor S, Mohamed RN, Shukri NA, Mahmoor MSN, Azillah N and Zamri F (2019) Overview on the theoretical prediction of shear resistance of steel fibre in reinforced concrete beams. *IOP Conference Series: Earth and Environmental Science* **220**: 012033.
38. Abad BF, Lantsoght E and Yang Y (2019) Shear capacity of steel fibre reinforced concrete beams. *Proceedings of the FIB Symposium 2019: Concrete – Innovations in Materials, Design and Structures* (Derkowski W, Krajewski P, Gwozdziewicz P, Pantak M, Hoidys L, (eds)). International Federation for Structural Concrete (fib), pp. 1–8.

39. Ananthi GBG, Sathick AJ and Abirami M (2020) Experimental investigation on shear behaviour of fibre reinforced concrete beams using steel fibres. *Journal of Construction Materials* **2**: 1–5.
40. Krassowska J and Kazberuk MK (2021) The effect of steel and basalt fibers on the shear behavior of double-span fiber reinforced concrete beams. *Materials MDPI* **14**: 6090.
41. Ng TS, Htut T and Foster SJ (2012) Fracture of steel fibre reinforced concrete – the unified variable engagement model. *UNICIV REPORT No. R-460*. The University of New South Wales, Sydney, Australia.
42. Bitencourt LAG, Manzoli OL, Bittencourt TN and Vecchio FJ (2019) Numerical modeling of steel fiber reinforced concrete with a discrete and explicit representation of steel fibers. *International Journal of Solids and Structures* **159**: 171–190.
43. BS EN 1992-1-1 (2004) Eurocode 2: Design of concrete structures – Part 1-1: General rules and rules for buildings. BSI, London, UK.
44. Narayanan R, Darwish IYS (1987) Punching shear tests on steel-fibre-reinforced micro-concrete slabs. *Magazine of Concrete Research* **39(138)**: 42–50.
45. Higashiyama H, Ota A and Mizukoshi M (2011) Design equation for punching capacity of SFRC slabs. *International Journal of Concrete Structures and Materials* **5(1)**: 35–42.
46. Maya LF, Fernández Ruiz M, Muttoni A and Foster SJ (2012) Punching shear strength of steel fibre reinforced concrete slabs. *Engineering Structures* **40**: 83–94.
47. Muttoni A and Fernández Ruiz M (2010) MC2010: the critical shear crack theory as a mechanical model for punching shear design and its application to code provisions. *FIB Bulletin 57: Shear and Punching Shear in RC and FRC Elements*, Lausanne (Switzerland): 31–60.
48. Frazão C, Camões A, Barros J and Gonçalves D (2015) Durability of steel fiber reinforced self-compacting concrete. *Construction and Building Materials* **80**: 155–166.
49. Alsaif A, Bernal SA, Guadagnini M and Pilakoutas K (2018) Durability of steel fiber reinforced rubberised concrete exposed to chlorides. *Construction and Building Materials* **188**: 130–142.
50. Meson VM (2019) *Durability of Steel Fibre Reinforced Concrete in Corrosive Environments*. PhD thesis, Technical University of Denmark, Denmark.
51. Chandra Paul S, van Zil GPAG and Šavija B (2020) Effect of fibers on durability of concrete: a practical review. *Materials MDPI* **13**: 4562.
52. Marcos-Meson V, Fischer G, Edvardsen C, Skovhus TL and Michel A (2019) Durability of steel fibre reinforced concrete (SFRC) exposed to acid attack – a literature review. *Construction and Building Materials* **200**: 490–501.

Fibre-reinforced Concretes for High-performance Structures: Building a more sustainable future

Lampropoulos A
ISBN 978-0-7277-6556-7
https://doi.org/10.1680/fchs.65567.031

Chapter 2
Ultra-high-performance fibre-reinforced concrete (UHPFRC)

2.1. Introduction and historical developments

Ultra-high-performance fibre-reinforced concrete (HPFRC) is a relatively new type of FRC with superior characteristics and significantly enhanced compressive and tensile strength. These properties are attributed to its special mix design and, more specifically, to the low water to cement ratio, the high amounts of cement and ground granulated blast furnace slag (GGBS) and the packing of the aggregates, combined with a high volume of steel fibres (2–6%).

Concrete technology has developed over the years since the early nineteenth century when Portland cement was first produced in England. Later, steel bars were added to concrete and reinforced concrete (RC) was developed. In the 1980s, high-performance concrete (HPC) was developed with supplementary materials (for example, silica fume), high-strength cement and low water to strength ratio, which led to enhanced compressive strength above 55 MPa. Further development and enhancement of the mechanical properties has been achieved over the years and the term ultra-high-performance concrete (UHPC) was coined in 1994. This term is normally used for concrete with compressive strength around 150–200 MPa. The term UHPC is used for a wide range of different types of ultra-high-performance concretes, including those in which fibres are used; these are known as ultra-high-performance fibre-reinforced concrete (UHPFRC). The first application of UHPFRC was for a pedestrian bridge in Canada in 1997 (Figure 2.1).

Since then there has been a continuous development of UHPFRC aimed at the enhancement of the mechanical characteristics as well as the use of sustainable resources and a reduction in the cost. Applications include the construction of new structures or the strengthening of existing ones.

Often, in addition to these superior properties, the concrete is required to be self-compacting. UHPFRC combines the properties of self-compacting concrete (SCC) with those of HPC and FRC (Figure 2.2).

The characteristics of UHPFRC are attributed to the special mix design that has a high amount of binder combination using cement, GGBS and silica fume, low water to cement ratio, superplasticiser, and a high percentage of steel fibres, which are normally in the range of 2–6%. A comparison of typical mixes for UHPFRC and normal concrete (NC) is presented in Figure 2.3.

The mix design is dependent on the requirements. The compressive strength characteristics are mostly affected by the cementitious matrix, whereas the tensile strength characteristics are affected by the amount and type of the steel fibres.

Figure 2.1 History of UHPFRC development (UHPC solutions)

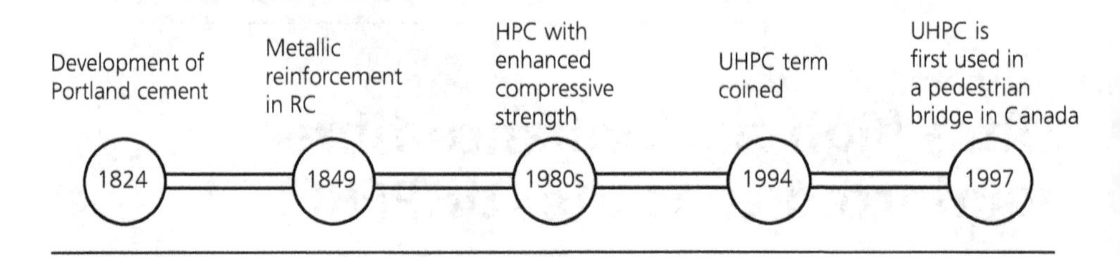

Figure 2.2 UHPFRC has a combination of the characteristics of SCC, FRC and HPC [1]

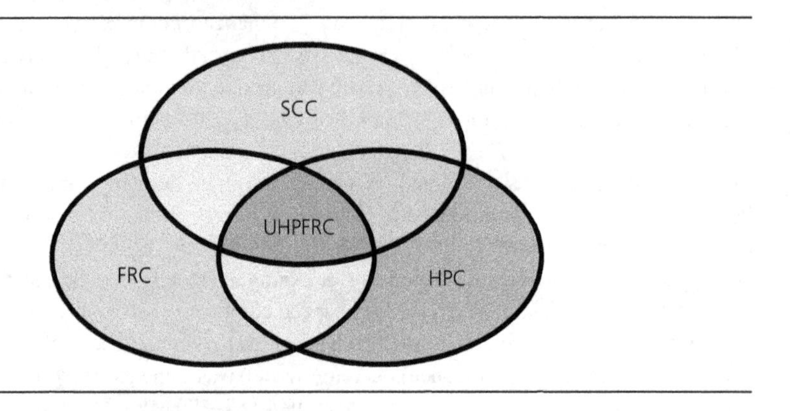

Figure 2.3 Typical mixes for UHPFRC and NC

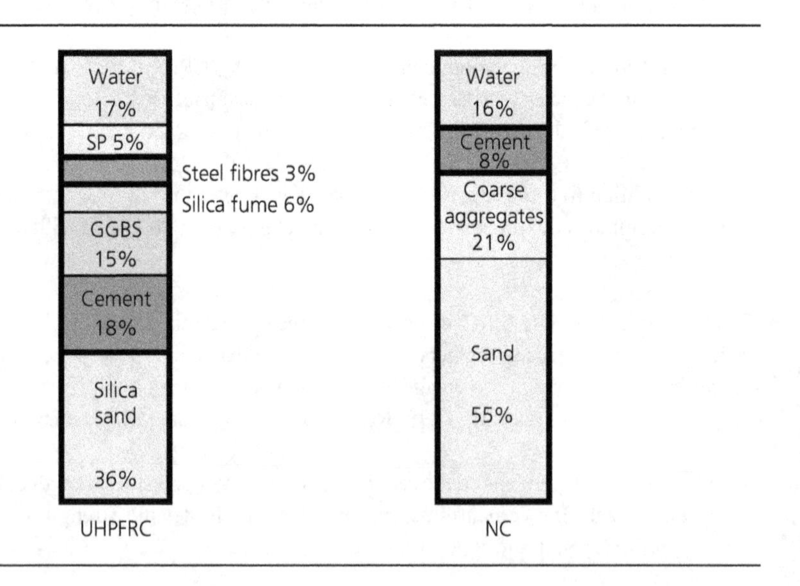

2.2. Materials selection

2.2.1 Cement

In most cases, high-strength cement (52.5) is used to provide high compressive strength and mechanical characteristics. Comparisons were made with lower-strength concrete and the results show that the mechanical characteristics are significantly affected by the cement [2]. More specifically, UHPFRC mixes were examined with 32.5 R type II and 52.5 N type I cement and the results show that when 52.5 N type was used as a replacement of 32.5 R cement, the compressive strength was increased by 16% and the tensile strength was 18% higher [2]. The most commonly used cement types are Portland cements (CEM I) and Portland slag cements (CEM II) of category 42.5 and 52.5 with normal (N) and rapid (R) hardening [3]. Experimental investigations have been conducted [3] and the results have shown that the type of cement significantly affects compressive strength and workability. In this study [3], two types of cement were classified as unsuitable for UHPFRC: CEM II 42.5 R and CEM I 42.5 N [3].

2.2.2 Aggregates

The role of the aggregates is of high importance for the mechanical performance of UHPFRC and, in general, fine materials should be used and the mixes should be designed in such a way as to provide the maximum possible particle packing to optimise the use of the binder and the mechanical properties of UHPFRC (Figure 2.4).

Research studies have shown the importance of particle packing on the mechanical performance of UHPFRC, and various models were proposed for the optimum design of UHPFRC mixes [4–10].

One of the most widely accepted approaches is the modified Andreasen and Andersen model, which is utilised to calculate the distribution of the solids.

More specifically, a target function is calculated using Equation 2.1 [6].

Figure 2.4 Indicative UHPFRC particle packing compared with conventional concrete

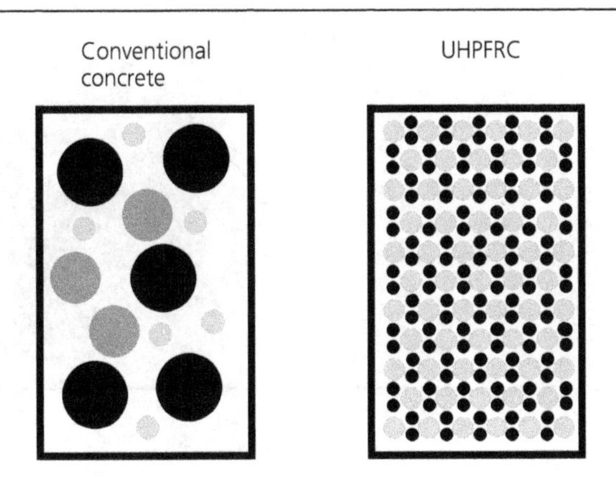

$$P(D) = \frac{D^q - D^q_{\min}}{D^q_{\max} - D^q_{\min}} \tag{2.1}$$

where:

$P(D)$ is the fraction of the total solids smaller than D
$D_{\max}$ is the maximum particle size (nm)
$D_{\min}$ is the minimum particle size (nm)
q is a coefficient for the distribution modulus which can be taken to be equal to 0.23 for mixes with high content of fine materials [6].

An optimisation process is required to adjust the proportion of each of the solid materials so as to minimise the deviation between the composed and the targeted curve (Figure 2.5). The deviation is expressed as the sum of the residual squares (RSS) which is taken at different particle sizes (D_i) taken at n points between $D_{\max}$ and $D_{\min}$ (Equation 2.2).

$$\text{RSS} = \frac{\sum_{i=1}^{n} \left(P_{\text{mix}}\left(D_i^{i+1}\right) - P_{\text{tar}}\left(D_i^{i+1}\right)\right)^2}{n} \tag{2.2}$$

where:

$D_{\min}$ is the examined mix
P_{tar} is the targeted mix using Equation (2.1).

2.2.3 Steel fibres

The presence of fibres is of fundamental importance for the behaviour of UHPFRC.

The most common approach includes the addition of fibres with 2–6% per volume, and normally steel fibres are used. The volume fraction of fibres and the geometry of the steel fibres

Figure 2.5 Indicative particle distribution for the particle packing process

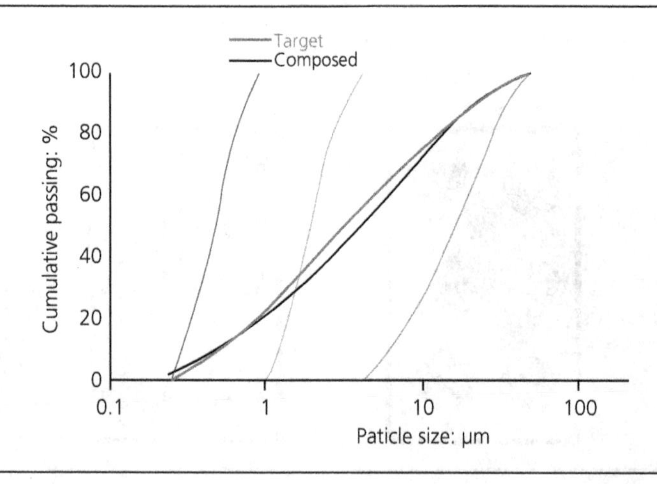

may significantly affect the compressive strength characteristics and play a crucial role in the flexural/tensile behaviour of UHPFRC. These are discussed in the following sections.

2.3. UHPFRC mechanical characteristics for varying fibre volume fractions

In this section, the main compressive and tensile strength characteristics of UHPFRC and the effect of the fibres on these properties will be discussed.

2.3.1 Compressive behaviour

The compressive strength is one of the main characteristics of UHPFRC and its superior properties are mostly attributed to the composition of the binder, which provides the main compressive strength. The contribution of the fibres is not expected to be of great importance. However, a beneficial effect is anticipated since the fibres restrain crack development, which provides additional strength and ductility. The beneficial effect of the fibres is affected by the geometry of the fibres and the volume fraction. Also, the presence of the fibres may have a detrimental effect due the fibre agglomeration or entrapped air, which may significantly affect the particle packing of the UHPFRC binder.

One of the most popular types offibre is straight steel microfibre with length 13 mm. The main benefit of these fibres is the fact that mixing can be facilitated, which limits the risk of agglomeration even for quite high percentages of volume fractions.

Additional fibre types were used such as hooked-end, corrugated, twisted and spiral but the general conclusion is that the straight microfibres have a beneficial effect owing to the bridging of the microcracks, whereas longer fibres with various shapes and improved pull-out performance do not offer significant improvement of the compressive strength characteristics since the favourable pull-out effect will not be utilised; inappropriate fibre distribution may also have a detrimental effect [11, 12].

Figure 2.6 shows an indication of the ratio of the compressive strength of UHPFRC to the respective strength for the same mix without fibres, for different volume fraction values for straight and corrugated fibres. This is a qualitative indication which is based on experimental results from the literature [11], and the actual effect of the type and volume fraction needs to be examined each time through detailed testing.

In general, the addition of microfibres leads to enhanced compressive strength characteristics which is attributed to the bridging of the microcracks. This strength increment may be quite important and may also reach 50% of the respective compressive strength of the mixes. However, when there is a high percentage offibres and when the distribution of the fibres is not uniform or when there are issues due to fibre agglomeration, the effect of the fibres may be detrimental and special attention is required.

2.3.2 Tensile stress–strain behaviour

The main contribution of the fibres in UHPFRC is on the tensile stress–strain characteristics. The contribution of fibres is of great importance after the end of the elastic part and, in particular, in the post-peak region whereas the elastic part of the stress–strain behaviour is not

Figure 2.6 Indicative compressive strength ratios for various fibre types and volume fractions (based on published data [11])

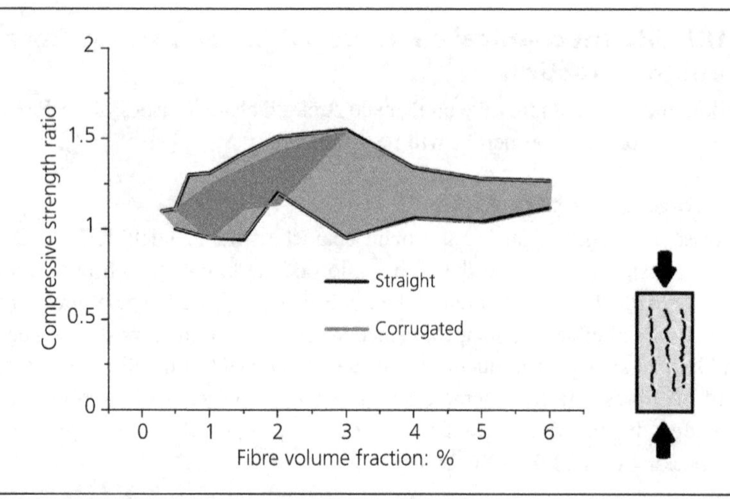

significantly affected by the presence of fibres. The actual tensile stress–strain behaviour of the UHPFRC can be determined through direct tensile tests (Figure 2.7).

These types of tests can offer valuable information about the stress–strain characteristics of the material in tension. However, particular attention needs to be paid to prevent the induction of moment through the grips of the specimens and to avoid stress localisation and failure of the

Figure 2.7 Direct tensile tests to determine the tensile stress–strain characteristics of UHPFRC

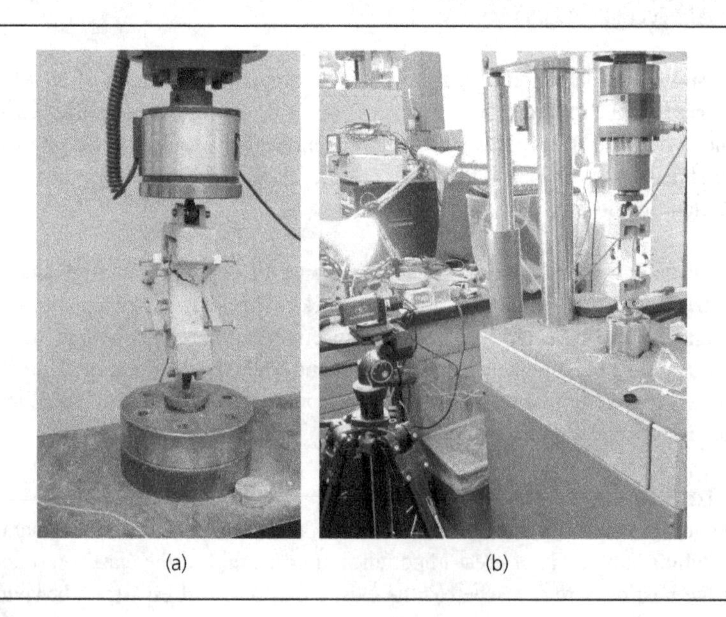

specimens at the ends of the grips. Also, the calculation of strains is a quite complicated process because, after the initial cracking of the specimens, there is a localisation of the failure around the cracks, and therefore it is more realistic after a certain level of damage to present the results in terms of 'crack opening' rather than using strain values.

The behaviour of UHPFRC in tension can be classified in three different stages (Figure 2.8).

Stage I is the uncracked stage, which is characterised by a linear stress–strain behaviour. Stage II is the strain-hardening stage where the fibres bridge the microcracks, and even if the stiffness at this stage is reduced, the stress capacity is further increased. In stage III, damage localisation occurs, and at this stage there is opening of the main crack and the stress capacity starts to reduce.

The strain hardening and softening characteristics of stages II and III are highly dependent on the type and volume fraction of the fibres. In general, a high percentage of fibres allows a more effective bridging of the cracks and, therefore, enhanced tensile strength characteristics and improved stress retention in the softening branch. If there is insufficient bridging of cracks either owing to the lack of sufficient fibres or to the type of fibres, stage II will not exist, and a softening branch will follow the end of the elastic part without any strain hardening.

Three different tensile stress–strain behaviours are proposed in the literature [2, 13]. Stress–strain curve without strain hardening (Figure 2.9(a)), with low strain hardening (Figure 2.9(b)) and with high strain hardening (Figure 2.9(c)).

In the case of stress–strain behaviour without strain hardening (Figure 2.9(a)), the tensile stress drops immediately after the end of the elastic part and when the cracks initiate without any contribution of the steel fibres. This is a typical behaviour for UHPFRC with relatively low fibre volume fractions, normally below 2%. When a higher percentage of fibres is used (for example, 2–3%), it is normal to have a strain-hardening branch and low strain-hardening

Figure 2.8 Characteristic stages of typical UHPFRC tensile stress–strain behaviour

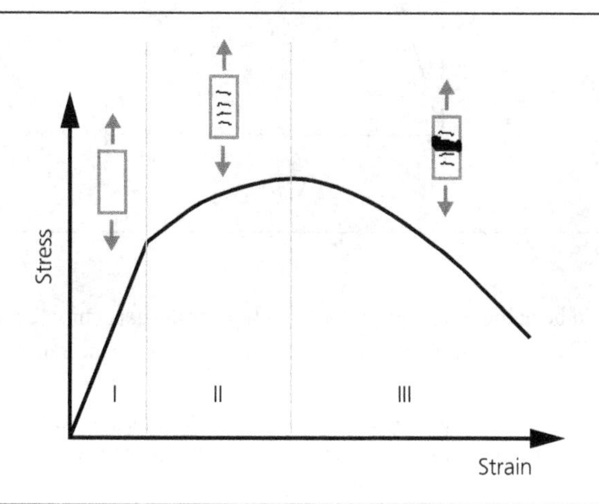

Figure 2.9 Indicative UHPFRC tensile stress–strain behaviour (a) without strain hardening, (b) with low strain hardening and (c) with high strain hardening [2]

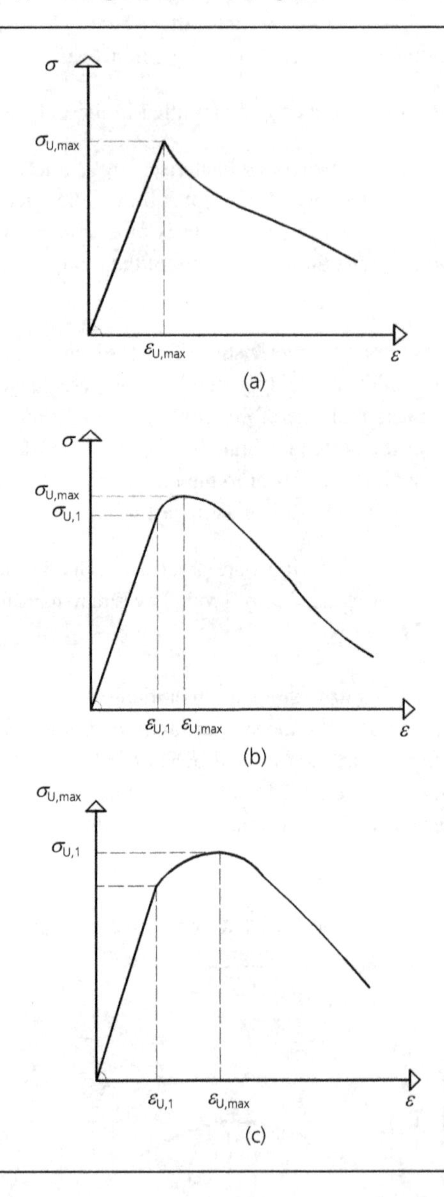

behaviour is likely to be achieved (Figure 2.9(b)). High strain-hardening behaviour is normally achieved for UHPFRC with a high volume fraction normally in the range of 4–6% [2].

The results of a previous published study [2] were used to give indicative values of the characteristic points of stress and strain for the three different percentages of steel fibres following the proposed simplified models presented in Figure 2.10. Here, $\sigma_{u,1}$ and $\varepsilon_{u,1}$are the stress and strain at the end of the elastic part of the graph, $\sigma_{u,2}$ and $\varepsilon_{u,2}$ are the stress and strain at the end of the second part of the graph (in the case of strain hardening), and $\sigma_{u,max}$ and $\varepsilon_{u,max}$ are the respective values at the ultimate stress value. E_u is the modulus of elasticity and $E_{U,hard}$ is the slope of the second linear part of the graph in the case of strain hardening.

Figure 2.10 Proposed simplified models for UHPFRC tensile stress–strain behaviour (a) without strain hardening, (b) with low strain hardening and (c) with high strain hardening [2]

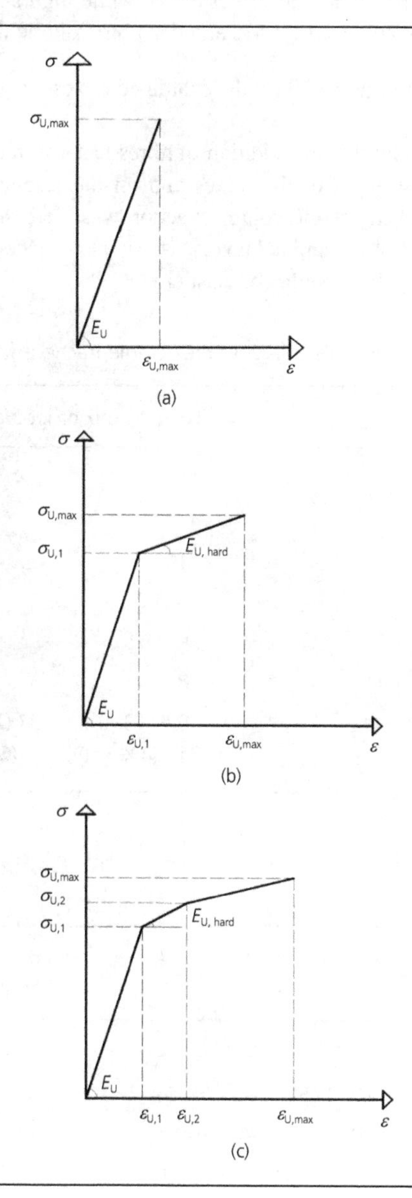

Extensive experimental work was conducted and direct tensile tests were executed using the set-up of Figure 2.7.

The examined mix design is presented in Table 2.1 [2].

The direct tensile tests show that the model of Figure 2.10(a) is representative of the behaviour of UHPFRC with 1% steel fibres, the model of Figure 2.10(b) is representative of the behaviour of UHPFRC with 2% and 3% steel fibres, while high strain-hardening characteristics (Figure 2.10(c)) were achieved for 4% and 6% fibre volume fractions.

The characteristic points of Figure 2.10 for the examined mixes are presented in Table 2.2 [2].

The results of Table 2.2 show that the addition of fibres has a significant beneficial effect on both the ultimate tensile strength of the mixes and on the respective values of the strain capacity which were increasing as the volume fraction was increasing. The Young modulus (E_u) was quite similar for all the examined mixes, as expected, since the presence of the fibres does not significantly affect the modulus of elasticity.

Table 2.1 UHPFRC mixed design with different fibre volume fractions [2]

Material	UHPFRC mix proportions: kg/m^3
Cement	657
GGBS	418
Silica fume	119
Silica sand	1051
Superplasticisers	59
Water	185
Steel fibres	78.5 (1% vol.), 157 (2% vol.), 235.5 (3% vol.), 314 (4% vol.), 471 (6% vol.)

Table 2.2 Characteristic points of the stress–strain models (Figure 2.10) for mixes with various fibre contents [2]

Fibre volume fraction: %	$\sigma_{u,1}$: MPa	$\varepsilon_{u,1}$	$\sigma_{u,2}$: MPa	$\varepsilon_{u,2}$	$\sigma_{u,max}$: MPa	$\varepsilon_{u,max}$	E_u: GPa
1	–	–	–	–	6.5	0.000 12	52.4
2	8.0	0.000 15	–	–	8.4	0.000 36	53.3
3	8.9	0.000 17	–	–	9.6	0.000 50	52.3
4	9.1	0.000 17	10.6	0.000 27	11.3	0.000 55	53.5
6	8.5	0.000 16	11.5	0.000 30	12.5	0.001 20	53.1

2.3.3 Shear behaviour

The shear performance of UHPFRC structural elements is affected by the contribution of concrete ($V_{Rd,c}$), the contribution of any available shear links ($V_{Rd,s}$) and the contribution of fibres ($V_{Rd,f}$) [14] (Equation 2.3).

$$V_{Rd} = V_{Rd,c} + V_{Rd,s} + V_{Rd,f} \tag{2.3}$$

The calculation of the contribution of concrete ($V_{Rd,c}$) can be carried out using Equation 2.4 [14].

$$V_{Rd,c} = \left[C_{Rd,c} \times k \times (1000 \times \rho_1 \times f_{ck})^{1/3} + k_1 \times \sigma_{cp}\right] \times b_w \times d \tag{2.4}$$

f_{ck} and σ_{cp} are in MPa

where:

$C_{Rd,c} = 0.15/\gamma_c$ where γ_c is concrete safety factor
$k \quad = 1 + \sqrt{200/d} \leq 2.0 \quad (d \text{ in mm})$
$k_1 \quad = 0.12$
$\sigma_{cp} \quad = N_{Ed}/A_c < 0.2 \times f_{cd}$
b_w is the smallest width of the cross-section in the tensile part
d is the effective depth of the cross-section
ρ_1 is the reinforcement ratio of the longitudinal reinforcement (A_s)
$= A_s/(b_w \times d) \leq 0.06 \quad (d \text{ in mm})$
N_{Ed} is the axial force
A_c is the concrete cross-sectional area.

For conventional concrete, ρ_1 is limited to 2%, but for UHPFRC the upper limit of ρ_1 is increased to 6% [14].

For the shear reinforcement, Equation 2.5 can be used to calculate the contribution of the shear links located at a spacing (s) ($V_{Rd,s}$) [14].

$$V_{Rd,s} = \frac{A_{sw}}{s} \times z \times f_{ywd} \times (\cot\theta + \cot\alpha) \times \sin\alpha \tag{2.5}$$

where:

$z = 0.9 \times d$, which should be no longer than $z = d - 2 \times c_{V,l} \geq d - c_{V,l} - 30$ mm
A_{sw} is the shear reinforcement cross-sectional area
f_{ywd} is the design yield strength of the shear reinforcement
α is the angle between the shear reinforcement and the axis of the structural element
$c_{V,l}$ is the compressive reinforcement concrete cover.

In this case, and since $V_{Rd,c}$ and $V_{Rd,s}$ are both taken into consideration, the usual limits proposed for angle θ are not applicable and the following equation (Equation 2.6) is proposed [14].

$$\cot\theta = 1.2 + 2.4 \times \frac{\sigma_{cp}}{f_{cd}} \tag{2.6}$$

where σ_{cp} is the longitudinal stress.

The shear resistance of the fibres ($V_{Rd,f}$) is given by Equation 2.7 [14].

$$V_{Rd,f} = b_w \times h \times n_f \times f_{cftd} \tag{2.7}$$

where:

$f_{cftd} = \alpha_{cf} \times f_{cftk}/\gamma_{CF}$
$f_{cftk} = \kappa_f \times f_{cft0}$
h is the height of the cross-section
n_f is a coefficient that considers the shape of the shear crack and is taken to be equal to 1 for I shaped cross-sections and 0.7 for rectangular cross-sections
a_{cf} is a coefficient for the long-term effects on post-cracking tensile strength
κ_f is a coefficient that considers the fibre orientation and is taken to be equal to 1 for I shaped cross-sections and 0.5 for rectangular cross-sections
γ_{CF} is the partial factor for the post-cracking tensile strength [14].

The basic value of the post-cracking tensile strength (f_{cft0}) is normally derived from material testing, but, in the absence of experimental tests, the following equation is proposed to give an approximate value (Equation 2.8) [14].

$$f_{cft0} = 0.07 \times \rho_f \times \frac{l_f}{\varphi_f} \tag{2.8}$$

ρ_f is the fibre volume fraction
l_f is the length of the fibres
φ_f is the diameter of the fibres.

2.3.4 Punching shear of UHPFRC slabs

UHPFRC is a relatively new construction material and the punching shear performance of UHPFRC has recently been studied [15–17].

Extensive experimental work was conducted to evaluate the performance of different types of UHPFRC slabs [15]. The application of Equation 2.9 was considered for calculating the punching shear resistance (V_u) of the UHPFRC slabs that were examined experimentally.

$$V_u = v_u \times b_o \times d \tag{2.9}$$

where:

v_u is the punching shear stress
b_o is the basic control perimeter
d is the thickness of the slabs.

The basic control perimeter (b_o) can be calculated using Equation 2.10.

$$b_o = 2 \times \pi \times \left(2.5 \times d + \frac{\text{diam}_{\text{load}}}{2}\right) \tag{2.10}$$

where:

$\text{diam}_{\text{load}}$ is the diameter of the loading point.

The proposed value of 2.5*d* for the critical control perimeter is based on experimental results [15]. It needs to be mentioned that existing code recommendations propose smaller values (varying from 0.5*d* to 2*d*) for design purposes.

Values for the ultimate punching shear stress (v_u) were derived experimentally for different percentages of steel fibre with and without the presence of longitudinal steel bars.

The following figure (Figure 2.11) shows the calculated ultimate punching shear stress (v_u) for the different types of UHPFRC using different fibre volume fractions and different volumetric amounts of longitudinal reinforcement.

Figure 2.11 Punching shear strength (v_u) for different types of UHPFRC [15]

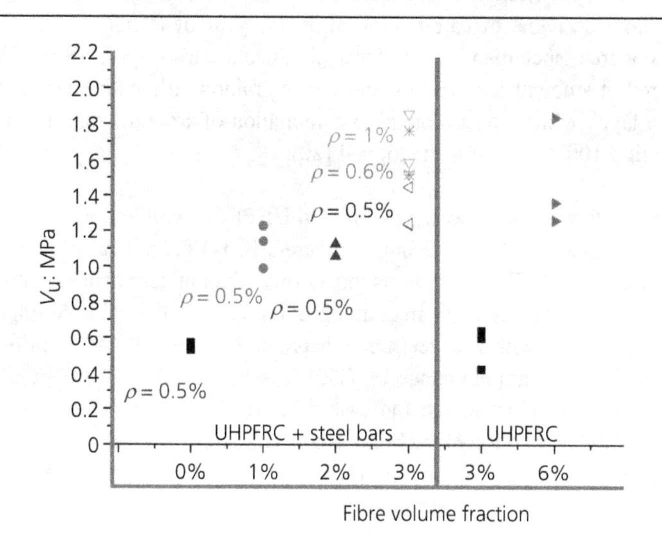

The results of Figure 2.11 give an indication of the shear strength values for UHPFRC slabs. Further work in this direction is required in addition to the development of a model based on mechanics.

2.3.5 Durability

The durability of UHPFRC is significantly enhanced compared with conventional concrete owing to its composition and, more specifically, the low water to binder ratio and the presence of fine aggregates [18].

With regard to water permeability, it was found that UHPFRC shows superior permeability resistance owing to its dense microstructure. Experimental results show that a typical UHPFRC with 200 MPa compressive strength has a water absorption coefficient after 7.5 h equal to 0.2 kg/m^2 which is much lower than the respective value for conventional concrete C30 which was found to be equal to 2.7 kg/m^2 [18].

One of the most important parameters for the permeability of concrete is the water to binder (w/b) ratio. According to experimental studies in this field [18, 19], the 14 day permeability coefficient of UHPC with w/b ratio 0.4 was found to be equal to 0.04 mm^2/s, and this value was significantly reduced to 0.0025 mm^2/s for a w/b ratio equal to 0.17. The addition of nanoparticles to UHPC has also been found to lead to 36% less water absorption [18].

The corrosion resistance of UHPFRC with different percentages of steel fibres has also been studied [20]. Various curing conditions were examined and durability tests were performed to evaluate the corrosion resistance of UHPFRC [20]. More specifically, accelerated carbonation tests, chloride migration and chloride diffusion tests were performed and comparisons were made with a conventional concrete (C50).

The results show that the UHPFRC specimens had significantly enhanced resistance to carbonation, because all the examined UHPFRC specimens, apart from the one cured at ambient temperature, show no signs of carbonation after one year of testing. On the contrary, the conventional concrete specimens (C50) show significant carbonation depth. The UHPFRC specimens cured at ambient temperature show some minor carbonation signs, due to a thin porous surface layer, which corresponded to carbonation of around 1 mm depth. This developed over the first 100 days and then stopped [20].

The chloride migration has also been examined in UHPFRC with various fibre contents, and comparisons were made with conventional concrete (C50) [20]. The results show that the chloride migration in UHPFRC mixes was much lower than in conventional concrete and the difference was around two orders of magnitude. For example, the chloride migration coefficient value for UHPFRC with 2% steel fibres was found to be 2.0×10^{-13} m^2/s whereas the respective value for conventional concrete (C50) was equal to 2.1×10^{-11} m^2/s. The significantly reduced chloride migration in the case of UHPFRC is attributed to the very dense structure of UHPFRC in addition to the high cement content, which leads to the presence of an increased amount of tricalcium aluminate [20]. The presence of steel fibres also has a significant effect on the resistance to chloride penetration. More specifically, it was found that the presence of these fibres has both a favourable and an unfavourable effect. The favourable effect is linked to the control of shrinkage cracking, whereas the unfavourable effect is linked to the presence of the interphases among the materials and the subsequent porous zones [20].

2.4. Constitutive models for the design of UHPFRC elements

In this section, selected design models from SETRA, AFGC 2002 [13] and recommendations for the behaviour of UHPFRC in compression and tension are discussed.

The design models proposed for both tension and compression for strain-hardening and for strain-softening behaviours are presented in Figures 2.12(a) and 2.12(b), respectively.

Figure 2.12 UHPFRC design models for (a) strain-hardening and (b) strain-softening behaviour [13]

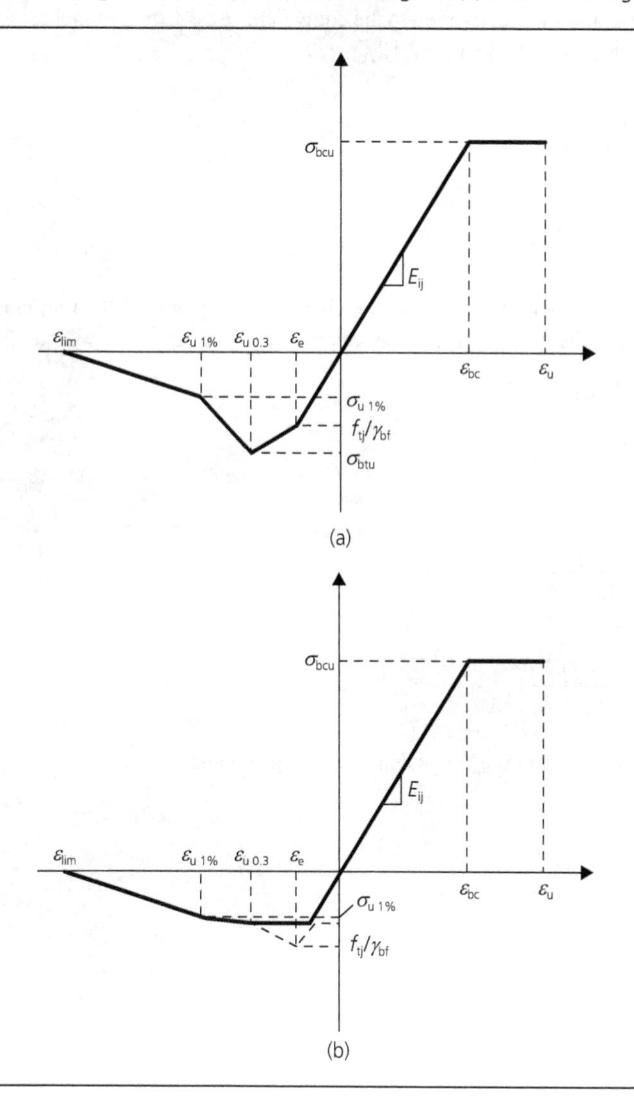

For the behaviour in compression, the ultimate stress (σ_{bcu}) is calculated using Equation 2.11.

$$\sigma_{bcu} = 0.85 \times f_{ck}/\gamma_{bf} \tag{2.11}$$

where:

f_{ck} is the characteristic compressive strength of UHPFRC
γ_{bf} is a safety factor that is taken to be equal to 1.3 for a fundamental load combination and 1.05 for an accidental.

For the tensile stress–strain behaviour, it is recommended to first derive the stress–crack-opening model from an inverse analysis and then, in order to apply the model in the tensile stress block, this should be converted to the equivalent stress–strain model using the following characteristic length (L_c) (Equation 2.12).

$$L_c = 2/3 \times h \tag{2.12}$$

where:

h is the cross-sectional height.

The elastic strain (ε_e) and the strains at the characteristic points of 0.3 mm crack width ($\varepsilon_{0.3}$) and 1% of beam height ($\varepsilon_{1\%}$) can be calculated using Equations 2.13, 2.14 and 2.15, respectively.

$$\varepsilon_e = \frac{f_{tj}}{E_c} \tag{2.13}$$

$$\varepsilon_{0.3} = \frac{w_{0.3}}{L_c} + \frac{f_{tj}}{\gamma_{bf} \times E_c} \tag{2.14}$$

$$\varepsilon_{1\%} = \frac{w_{1\%}}{L_c} + \frac{f_{tj}}{\gamma_{bf} \times E_c} \tag{2.15}$$

The ultimate tensile strain (ε_{lim}) is calculated using Equation 2.16.

$$\varepsilon_{lim} = \frac{L_f}{L_c} \tag{2.16}$$

where:

L_f is the fibre length.

The stress values at $w_{0.3}$ and $w_{1\%}$ can be calculated using Equations 2.17 and 2.18.

$$f_{\text{bt}} = \frac{f_{(w_{0.3})}}{K \times \gamma_{\text{bf}}} \tag{2.17}$$

$$f_{1\%} = \frac{f_{(w_{1\%})}}{K \times \gamma_{\text{bf}}} \tag{2.18}$$

where:

f_{bt} is the stress at 0.3 mm crack width
$f_{1\%}$ is the stress at 1% of cross-sectional height (h) crack width
K is a coefficient that considers the effect of fibre orientation; recommended values are 1 for placement methods validated from testing, 1.25 for loading without expected local effects and 1.75 for local effects.

REFERENCES

1. Camacho ET (2013) *Dosage Optimization and Bolted Connection for UHPFRC Ties*. PhD thesis, University of Valencia, Spain.
2. Paschalis S and Lampropoulos A (2017) Fiber content and curing time effect on the tensile characteristics of ultra high performance fiber reinforced concrete. *Structural Concrete Journal of the fib* **18**: 577–588.
3. Hadl P, Kim H and Nguyen VT (2016) Influence of cement type and type of aggregate on the fresh and hardened properties of UHPC and HPC. *Proceedings of the 1st International Interactive Symposium on UHPC, Des Moines, Iowa, USA.*
4. Larrard F and Sedran T (1994) Optimization of ultra-high-performance concrete by the use of a packing model. *Cement and Concrete Research* **24**: 997–1009.
5. Larrard F and Sedran T (2002) Mixture-proportioning of high-performance concrete. *Cement and Concrete Research* **32**: 1699–1704.
6. Yu R, Spiesz P and Brouwers HJH (2015) Development of ultra-high performance fibre reinforced concrete (UHPC): towards an efficient utilization of binders and fibres. *Construction and Building Materials* **79**: 273–282.
7. Wang X, Yu R, Shui Z, Zhao Z, Song Q, Yang B and Fan D (2018) Development of a novel cleaner construction product: ultra-high performance concrete incorporating lead-zinc tailings. *Journal of Cleaner Production* **196**: 172–182.
8. Arora A, Almujaddidi A, Kianmofrad F, Mobasher B and Neithalath N (2019) Material design of economical ultra-high performance concrete (UHPC) and evaluation of their properties. *Cement and Concrete Composites* **104**: 103346.
9. Arora A, Yao Y, Mobasher B and Neithalath N (2019) Fundamental insights into the compressive and flexural response of binder- and aggregate-optimized ultra-high performance concrete (UHPC). *Cement and Concrete Composites* **98**: 1–13.
10. Akeed M, Qaidi S, Ahmed H, Faraj R, Majeed S, Mohammed A, Emad W, Tayeh B and Azevedo A (2022) Ultra-high-performance fiber-reinforced concrete. Part V: mixture design, preparation, mixing, casting, and curing. *Case Studies in Construction Materials* **17**: e01363.
11. Larsen I and Thorstensen RT (2020) The influence of steel fibres on compressive and tensile strength of ultra high performance concrete: a review. *Construction and Building Materials* **256**: 119459.

12. Wen C, Zhang P, Wang J and Hu S (2022) Influence of fibers on the mechanical properties and durability of ultra-high-performance concrete: a review. *Journal of Building Engineering* **52**: 104370.
13. SETRA, AFGC (2002) *Béton fibrés à ultra-hautes performances, (Ultra high performance fibre reinforced concretes)*, recommandations provisoires, 152 p., France.
14. Metje K and Leutbecher T (2023) Verification of the shear resistance of UHPFRC beams – design method for the German DafStb guideline and database evaluation. *Engineering Structures* **277**: 115439.
15. Lampropoulos A, Tsioulou O, Mina A, Nicolaides D and Petrou MF (2013) Punching shear and flexural performance of ultra-high performance fibre reinforced concrete (UHPFRC) slabs. *Engineering Structures* **281**: 115808.
16. Al-Quraishi HAA (2014) *Punching Shear Behaviour of UHPC Flat Slabs*. PhD thesis, University of Kassel, Germany.
17. Harris DK, Roberts-Wollmann CL (2005) Characterization of the punching shear capacity of thin ultra-high performance concrete slabs. In: *Final Report. Virginia Transportation Research Council.*
18. Li J, Wu Z, Shi C, Yuan Q and Zhang Z (2020) Durability of ultra-high performance concrete – a review. *Construction and Building Materials* **255**: 119296.
19. Tam CM, Tam VW and Ng KM (2012) Assessing drying shrinkage and water permeability of reactive powder concrete produced in Hong Kong. *Construction and Building Materials* **26**: 79–89.
20. Valcuende M, Lliso-Ferrando JR, Ramón-Zamora JE and Soto J (2021) Corrosion resistance of ultra-high performance fibre- reinforced concrete. *Construction and Building Materials* **306**: 124914.

Lampropoulos A
ISBN 978-0-7277-6556-7
https://doi.org/10.1680/fchs.65567.049

Chapter 3
Geopolymer cement-free concrete

3.1. Introduction and historical developments

Geopolymer concrete represents one of the main 'environmentally friendly' alternatives to conventional concrete. The basic principle is to replace cement with waste materials rich in silicon and aluminium oxides combined with activating chemicals that initiate the reactions and the development of geopolymerisation products and subsequent strength development. The most commonly used waste materials are pulverized fly ash (PFA) and ground granulated blast furnace slag (GGBS), although the use of other materials such as metakaolin has been explored.

The first reported alkali-activated concrete was made in 1930 when Kuhl [1] examined the use of slag-based mixes activated with dry potash solutions. Further developments were made by Chassevent in 1937 [2], whereby dry potash and soda solution were used to evaluate slag reactivity. In 1940, Purdon [3] conducted large-scale experiments on cement-free binders using slag and lime. In 1981, Davidovits [4] presented the results of concrete with a blend of metakaolinite limestone and dolomite and he was the first to use the term geopolymer which since then has been widely used in the scientific community. The use of slag-based geopolymers was first presented by Krivenko in 1986 [5], and the use offly-ash-based cements was first presented by Palomo *et al.* in 1999 [6]. Since then, a lot of research has been conducted on the development of various types of cement-free alkali-activated binders. This is an area which is continuously under development and the use of different sources of silicon and aluminium oxides and different activators have been examined [7–13].

The majority of these studies focused on the development of new types of alkali-activated cement and, in many cases, heat treatment was required to enable the activation process.

The use of fly ash (FA) in combination with GGBS was explored in various studies where different PFA to GGBS ratios were tested [14–19], and a heat-curing process was applied with quite promising results in terms of mechanical performance. Further research was made in this direction for the development of suitable geopolymer mixes cured under ambient temperature, which makes them suitable for in situ structural applications [20]. Also, further research was made in the enhancement of the mechanical characteristics and, specifically, the development of geopolymers with high performance and energy absorption with fibre reinforcement [21–24]. These materials have proved to be quite efficient for building new structures and also for the repair and strengthening of existing ones because, in addition to the enhanced properties, they are characterised by improved durability and they can protect the reinforcing steel bars from corrosion [21–24].

3.2. Materials selection

Various materials have been successfully used in the development of geopolymer concretes.

3.2.1 Raw materials/binders

The use of FA in combination with GGBS was extensively explored and can be successfully used for the development of geopolymer concrete with suitable characteristics [14–24]. There are different types of PFA and GGBS with various mix compositions and different degrees of fineness. These are parameters that may have a significant effect on the characteristics of the geopolymer concrete. Typical scanning electron microscope (SEM) images of PFA and GGBS are presented in Figures 3.1 and 3.2, respectively, where it can be seen that PFA consists mostly of spherical particles whereas GGBS particles are angular [20].

For PFA, it is important to use low calcium materials following the requirements of British Standards [25, 26] (Table 3.1).

With regard to fineness, coarser FA is classified as class N and finer FA is classified as class S. These are defined according to the percentage of the total sample mass that remains in a 45 μm sieve after sieve analysis; the values are less than 40% for class N and less than 12% for class S.

The finer the FA, the higher the surface area where geopolymerisation takes place and therefore the higher the reactivity of the material [25]. It is, therefore, preferable to use fine FA (that is, class S).

A combination of FA with GGBS is essential for the development of early strength under ambient temperature.

Figure 3.1 Typical PFA SEM image [20]

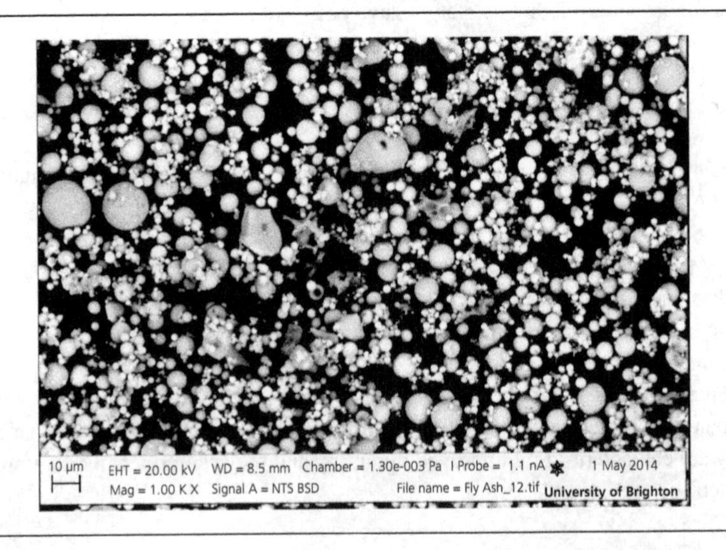

Figure 3.2 Typical GGBS SEM image [20]

Table 3.1 Chemical composition requirement for FA in concrete [25, 26]

Oxides	SiO_2 + Al_2O_3 + Fe_2O_3	SiO_2	CaO	Na_2O + K_2O	MgO	SO_3	Cl^-	P_2O_5	Loss on ignition
wt.%	≥ 70	≤ 25	≤ 10	≤ 5	≤ 4	≤ 3	≤ 0.1	≤ 5	A: ≤ 5 B: ≤ 7 C: ≤ 9

3.2.2 Mix design process

In this section, a procedure is presented for the calculation of the appropriate mix design considering the use of FA and GGBS.

The following mix design process is recommended [27].

STEP 1: Specify compressive strength, initial setting time, slump class.
STEP 2: Determine FA/GGBS and w/s ratios (Figure 3.3).
STEP 3: Check initial setting time. If it is equal or greater than the required one (Figure 3.4) then continue to **STEP 4**. Otherwise, increase GGBS, calculate new FA/GGBS and w/s ratio and repeat **STEP 3**.
STEP 4: Determine paste amount.
STEP 5: Determine binder content (Figure 3.5).
STEP 6: Determine quantities of activators.
STEP 7: Determine amount of water.
STEP 8: Determine amount of aggregates.

Figure 3.3 Compressive strength against w/s ratio for various fly ash/GGBS ratios [27]

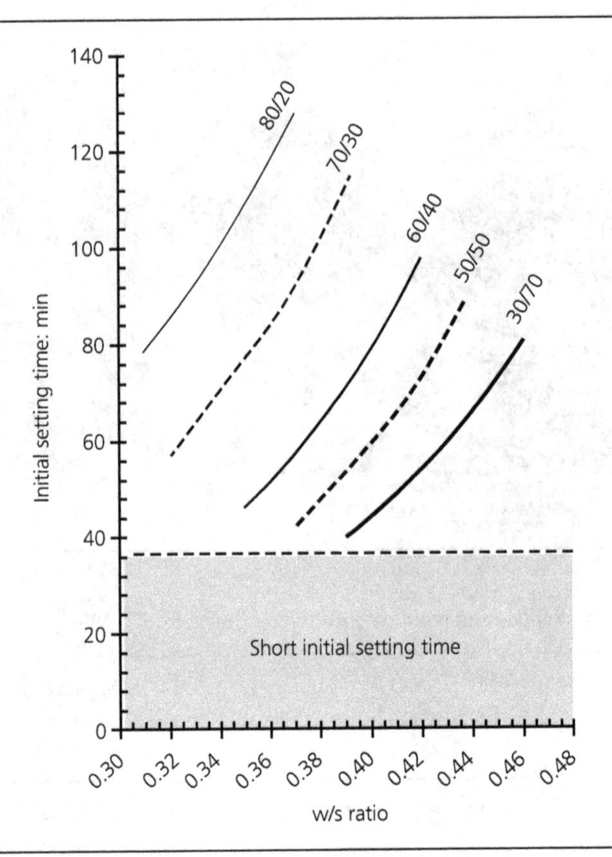

The proposed methodology starts with the selection of the appropriate strength setting time and slump (STEP 1). The compressive strength is used to determine the appropriate water to solids ratio (w/s) and the FA/GGBS ratio using Figure 3.3 (STEP 2).

The initial setting time of the examined mixes from Figure 3.4 should be checked with the required setting time (STEP 3). If the initial setting time is greater or equal to the required one, then the binder content needs to be calculated using Figure 3.5. Otherwise, and for lower values of the initial setting time, the amount of GGBS needs to be increased and STEP 2 should be repeated with calculation of increased w/s ratio which leads to increased setting time.

Then, in STEP 4, the amount of paste needs to be calculated. In the proposed methodology, a paste volume of 30–33% was examined and the exact amount is determined by the required consistency class; there are detailed graphs correlating the consistency with the paste volume [27]. If the required consistency can be achieved for various paste volumes, then the use of the lower value is recommended [27]. The binder quantity is calculated in STEP 5 using Figure 3.5.

Figure 3.4 Initial setting time against w/s ratio for various fly ash/GGBS ratios [27]

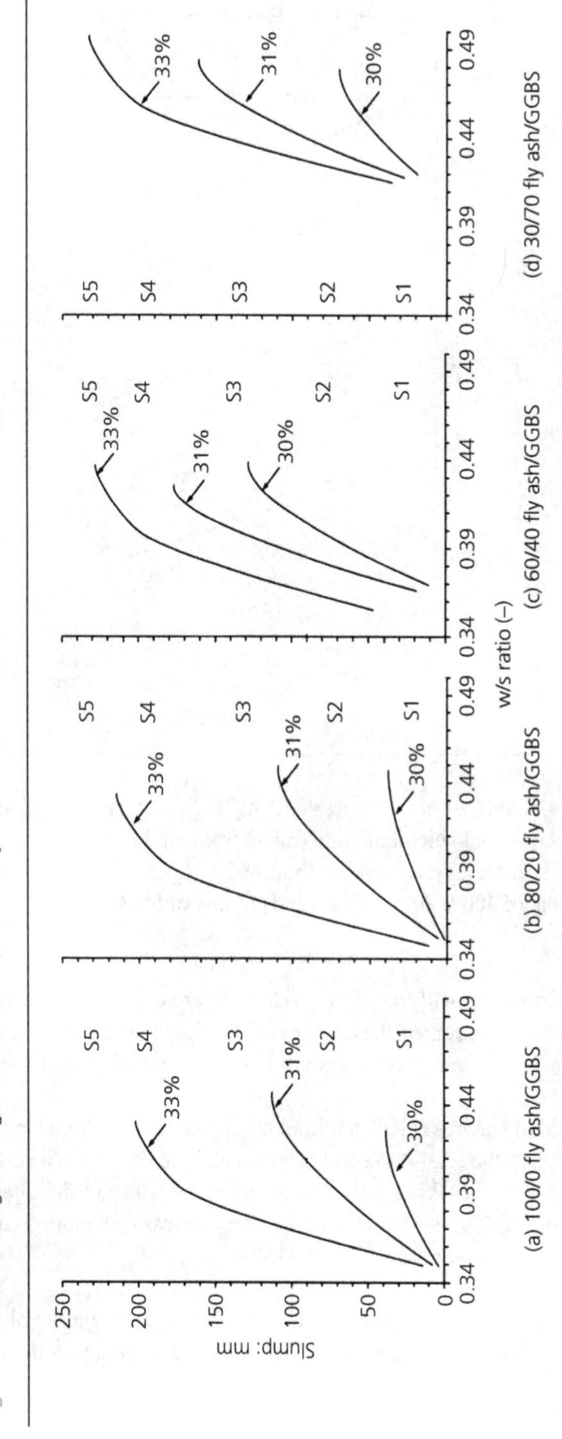

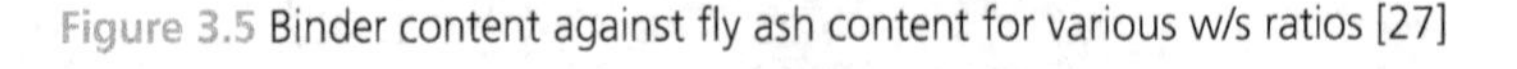

Figure 3.5 Binder content against fly ash content for various w/s ratios [27]

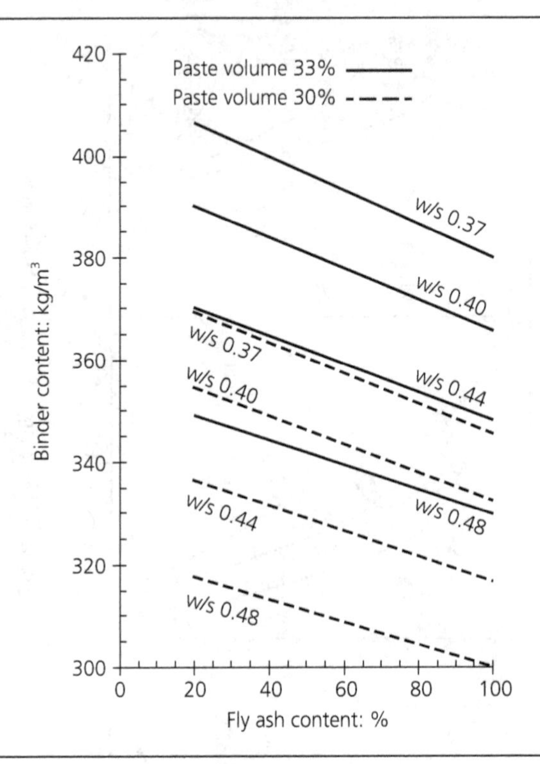

To determine the amounts of activators (STEP 6), it is proposed to calculate the amount of sodium oxide (Na_2O) and silicon dioxide (SiO_2) from the binder mass for M+ = 7.5% and AM = 1.25 where M+ = Na_2O/(FA + GGBS) and AM = Na_2O/SiO_2. For example, for a binder (FA + GGBS) content of 400 kg/m^3, the required amount of Na_2O is 30 kg/m^3 and the amount of SiO_2 is 24 kg/m^3.

The chemical activators normally used are sodium hydroxide (NaOH) and sodium silicate (Na_2SiO_3) and the composition of these chemicals varies depending on the providers. The Na_2O from both the NaOH and Na_2SiO_3 count towards the total required Na_2O.

The required amount of water is calculated from the specified w/s ratio where the solids (s) are considered as the sum of the FA, GGBS and alkali solids, and the water (w) is the total required amount of water in the mix taking into consideration the water in the chemical activators. Therefore, by subtracting the water included in the chemical activators from the total water (w), the additional water can be calculated (STEP 7).

The volume of aggregates (STEP 8) is calculated by subtracting the paste volume from 1 m^3 of concrete. The mass of each aggregate size is determined depending on the required strength and workability.

Figure 3.6 Upper and lower limits for the aggregate grading as proposed by DIN 1045 [29]

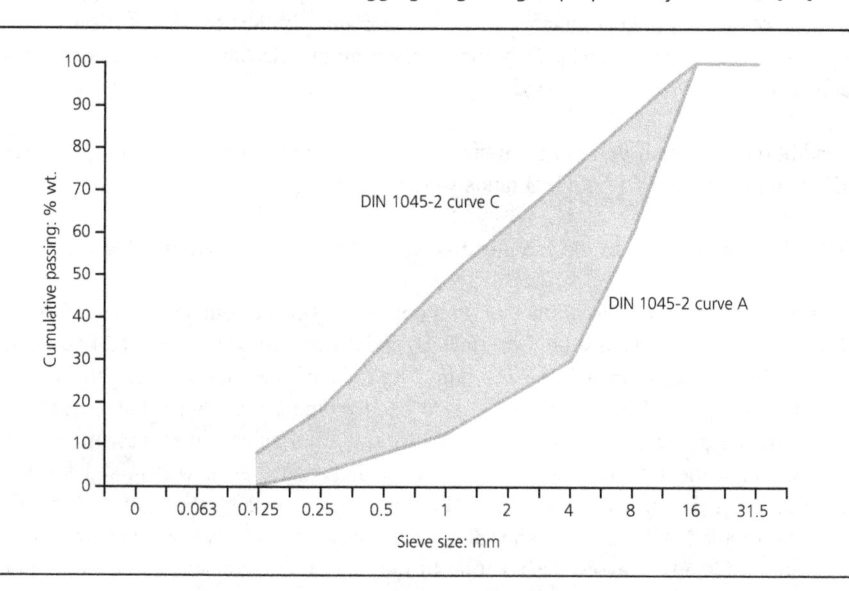

For high-strength concretes reinforced with fibres, the use of fine sand is recommended as this allows improved bonding between the fibres and the concrete matrix. For conventional applications, combined aggregate grading can be used in line with the limits proposed by DIN 1045 (Figure 3.6) [28, 29].

The exact type and grading of the aggregates depends on the requirements in terms of workability and water absorption. Research has been conducted in this field using three types of aggregates in geopolymer mortar namely, natural river sand, crushed limestone, and combined sand-limestone [30]. The results show that the combined sand-limestone with 50% river sand and 50% crushed limestone has reduced water absorption compared with the other type of aggregates. Also, the type of aggregate significantly affects the workability of the mix. Flow-test results show that the use of crushed limestone leads to reduced geopolymer mortar flow. The flow increases in the case of combined sand, and the use of natural sand shows further improvement in the flowability [30].

3.3. Fibre-reinforced geopolymer concrete

The addition of fibre reinforcement was found to be very efficient for the enhancement of the mechanical characteristics; specifically, the tensile stress–strain characteristics of geopolymer concrete [31, 32]. Extensive research was made on the selection of the appropriate geopolymer matrix and fibre type [31, 32].

3.3.1 Compressive behaviour and workability

The effect of the geopolymer matrix is of great importance for the behaviour offibre reinforced geopolymer concrete as this has a significant effect on the mechanical characteristics and on the fibre-to-geopolymer matrix bond.

The addition of a small percentage of silica fume was found to have a beneficial effect and various combinations of FA/GGBS ratios were examined [32].

The binder compositions of the examined geopolymers are presented in Table 3.2.

Potassium hydroxide was used together with potassium silicate solution as the activator. The potassium hydroxide was in solid form (pellets) and it was diluted with water 24 hours before mixing with the potassium silicate solution. The ratio of the pass of potassium hydroxide solution over the silicate solution was equal to 2.5, forming a solution modulus equal to 1.25. For the mixing process, the binder (FA, GGBS and solid silica fume) was first mixed for 5 minutes, then the activators and water were added, and then it was mixed for another 5 minutes. After this, the fibres were added to the mix followed by the sand, and the mixing continued for another 3 minutes. When slurry silica was used, this was added at the end (after the addition of sand) to avoid flash setting of the mix. The specimens remained at ambient temperature (21–23°C) until the day of the mechanical testing [32].

Table 3.2 Binder compositions of the examined geopolymer mixes [32]

Mixture ID	Slag: %	Silica fume type	Slag: kg/m^3	Silica fume: kg/m^3	Fly ash: kg/m^3
10S	10%	–	78	0	698
10S_10DSF		Densified	78	78	620
10S_10USF		Undensified	78	78	620
10S_55SSF		Slurry	78	39	659
20S	20%	–	155	0	620
20S-10DSF		Densified	155	78	543
20S_10USF		Undensified	155	78	543
20S_5SSF		Slurry	155	39	581
30S	30%	–	233	0	543
30S_0DSF		Densified	233	78	465
30S_10USF		Undensified	233	78	465
30S_5SSF		Slurry	233	39	504
40S	40%	–	310	0	465
40S_10DSF		Densified	310	78	388
40S_10USF		Undensified	310	78	388
40S_5SSF		Slurry	310	39	426

Flow-table tests were conducted (Figure 3.7) to evaluate the workability of the mixes. A conical mould was filled with the geopolymer mortar at two stages. Then the mould was removed and the table rose and dropped 25 times in a period of 15 seconds. The base diameter was measured at four almost equal intervals and the average diameter was used to evaluate the workability of the mixes.

The flow-table results are presented in Figure 3.8.

Figure 3.7 Flow-table test to evaluate workability

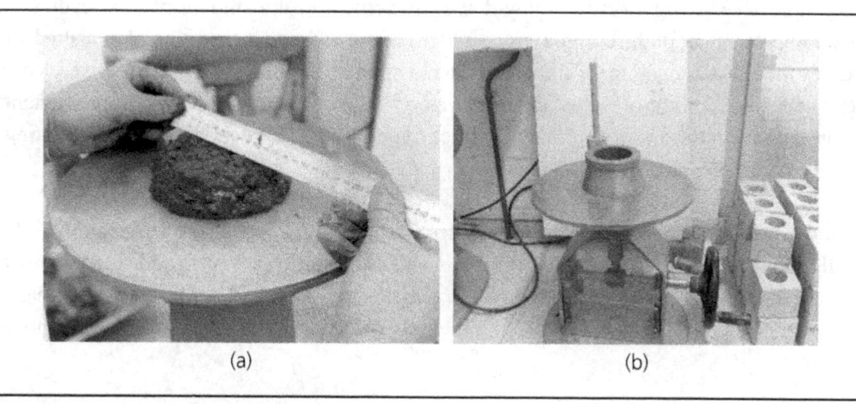

Figure 3.8 Flow-table test results for different types of binders

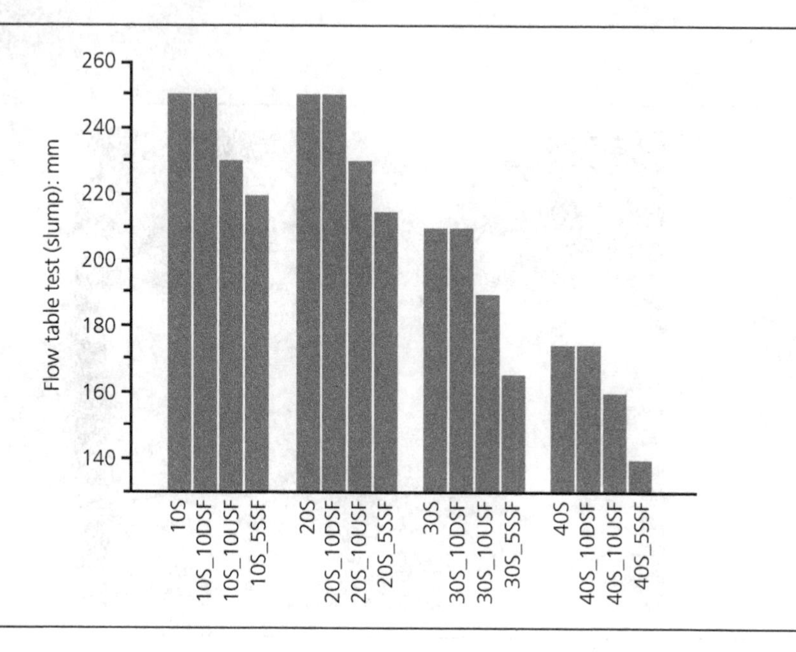

The results of Figure 3.8 show that the binder composition significantly affects the workability of the mixes. An increment in the slag content leads to significant reduction of the slump. This reduction is attributed to the increased reactivity of the slag. The addition of silica fume also leads to further reduction in the workability.

The mechanical characteristics of the examined mixes (Table 3.2) were evaluated 28 days after casting. Compressive strength was conducted on cubes of side 50 mm (Figure 3.9).

The results of the compressive tests are presented in Figure 3.10.

The compressive strength results show that an increment in the amount of slag leads to significant improvement of the compressive strength of the mixes. The addition of densified silica fume resulted in a reduction of the compressive strength in all the examined mixes. On the other hand, the addition of undensified and slurry silica fume results in an improvement in the compressive strength, with the undensified type being the most efficient and resulting in the highest compressive strength results.

The addition of fibre reinforcement to the mixes of Table 3.2 has also been examined. Specifically, straight steel fibres with length 13 mm and aspect ratio 81.5 were added to the mix at a volume fraction of 2%. The mix composition remains the same as that presented in Table 3.2. Also, the mix ID follows that of Table 3.2, the only difference being that '2 StF' is included at the end.

Compressive tests were conducted on cubes of side 50 mm (Figure 3.9) and the results are presented in Figure 3.11.

Figure 3.9 Compressive test on cubes of side 50 mm

Figure 3.10 Compressive test results of the mixes of Table 3.2

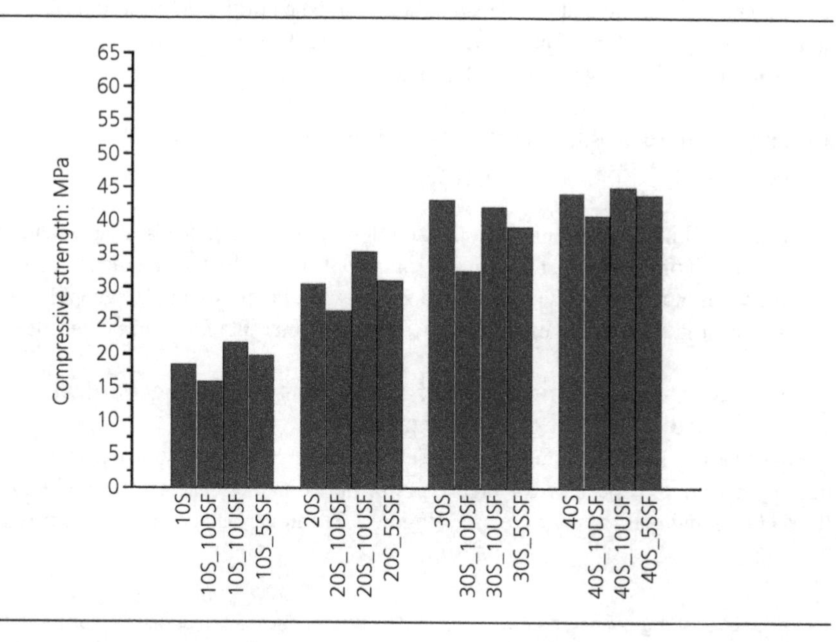

Figure 3.11 Compressive test results of steel fibre reinforced geopolymer mixes

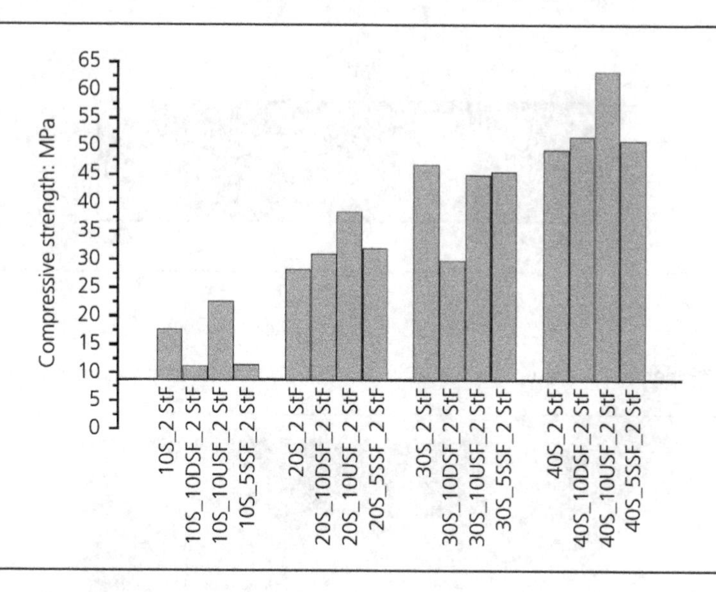

From the results of Figure 3.11 and by comparison with the respective values of Figure 3.10, it can be concluded that the addition of steel fibres does not significantly affect the compressive strength of geopolymer mixes with 10% slag. For higher slag percentages, the addition of steel fibres makes a significant contribution to the compressive strength compared with the respective values of the mixes without steel fibres.

3.3.2 Flexural performance and tensile stress–strain behaviour

To evaluate the flexural and tensile strength performance of the fibre-reinforced geopolymer concretes, flexural testing of prisms (Figures 3.12 and 3.13) and direct tensile testing of dog-bone shaped specimens were conducted (Figure 3.14).

The results of direct tensile tests of all the examined mixes with steel fibres are presented in Figure 3.15 [32].

The results of Figure 3.15 show that the addition of silica fume leads to an enhancement of tensile strength of the mixes. This is quite significant, especially when undensified silica fume is used, and is attributed to the enhanced bond between the fibres and the geopolymer matrix. An increment in the amount of slag also leads to a further increment in the tensile strength.

3.3.3 Evaluation of the effect of fibre types on the mechanical performance of geopolymers

Various types of fibres were studied for the enhancement of the mechanical performance of the geopolymer mixes. Specifically, three types of fibres, namely, steel (St), polyvinyl alcohol (PVA) fibres and glass fibres were used (Figure 1.3) and various percentages were examined.

Figure 3.12 Loading conditions for the testing of fibre-reinforced geopolymer concrete prisms

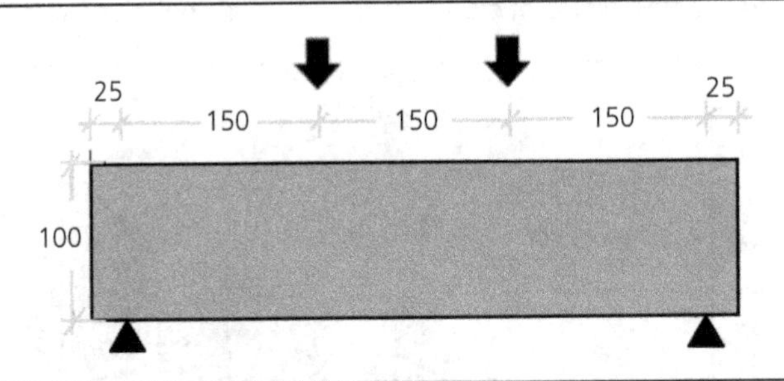

Figure 3.13 Experimental set-up for the testing of fibre-reinforced geopolymer concrete prisms

Figure 3.14 Direct tensile testing of fibre reinforced geopolymer concrete prisms

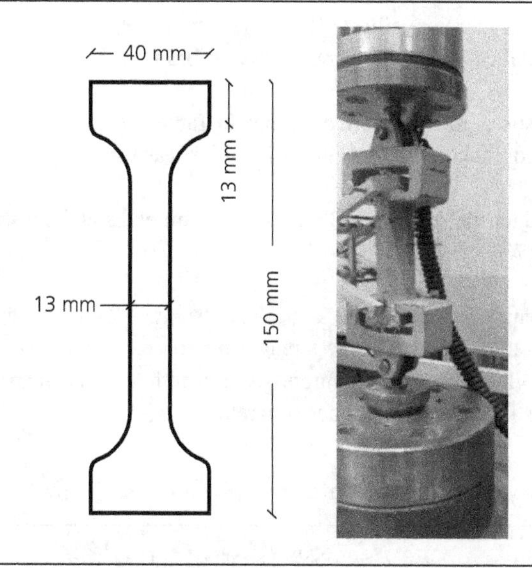

Figure 3.15 Tensile test results of steel-fibre-reinforced geopolymer mixes

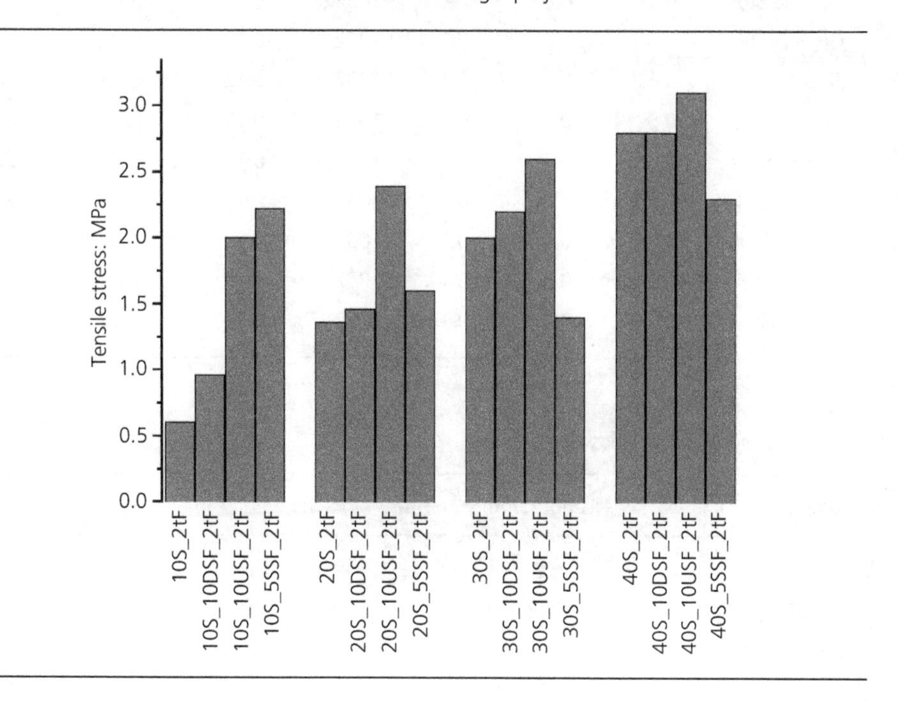

The mix design of the examined geopolymer is based on the previous mixes described in Table 3.2. More specifically, FA, GGBS and silica fume were used as a binder with a total amount of 775 kg/m^3. Potassium silicate (K_2SiO_3) with a molar ratio equal to 1.25 was used as the activator and silica sand with a total mass of 1054 kg/m^3 was also used (Table 3.3).

The exact characteristic of the fibres are described in Table 3.4. In one of the examined mixes (GP_3% St6/13), a combination of St6 and St13 fibres was used.

The compressive and tensile strength at 28 days was examined and the results are presented in Figures 3.16 and 3.17.

The compressive strength results (Figure 3.16) show that the addition of steel fibres leads to an enhancement of the 28-day compressive strength by approximately 10–20%. However, the addition of PVA and glass fibres show compressive strength results that are slightly lower than the respective values for plain geopolymer concrete.

Table 3.3 Binder compositions of the examined geopolymer mixes [31]

Mix ID	FA/ binder	GGBS/ binder	SF/ binder	Sand: kg/m^3	K_2SiO_3/ binder	Water/ binder	Fibre V_f: %
PGP	0.5	0.4	0.1	1052	0.12	0.25	0
GP_2% St6	0.5	0.4	0.1	1052	0.12	0.25	2
GP_3% St6	0.5	0.4	0.1	1052	0.12	0.25	3
GP_1% St13	0.5	0.4	0.1	1052	0.12	0.25	1
GP_2% St13	0.5	0.4	0.1	1052	0.12	0.25	2
GP_3% St13	0.5	0.4	0.1	1052	0.12	0.25	3
GP_3% St6/13	0.5	0.4	0.1	1052	0.12	0.25	3
GP_1% PVA	0.5	0.4	0.1	1052	0.12	0.25	1
GP_2% PVA	0.5	0.4	0.1	1052	0.12	0.25	2
GP_1% Glass	0.5	0.4	0.1	1052	0.12	0.25	1

Table 3.4 Fibre characteristics

Material	Length: mm	Diameter: mm	*E*: GPa
Steel (ST6)	6	0.16	200
Steel (ST13)	13	0.16	200
Glass	13	0.13	74
PVA	12	0.015	29.5

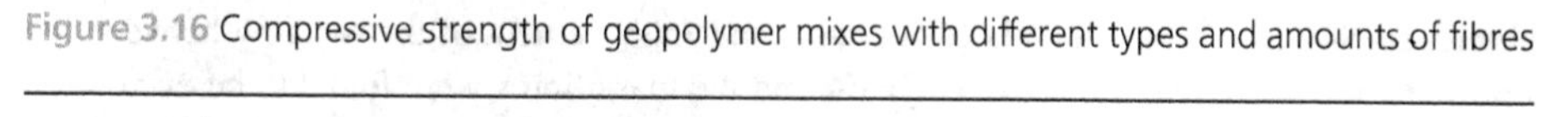

Figure 3.16 Compressive strength of geopolymer mixes with different types and amounts of fibres

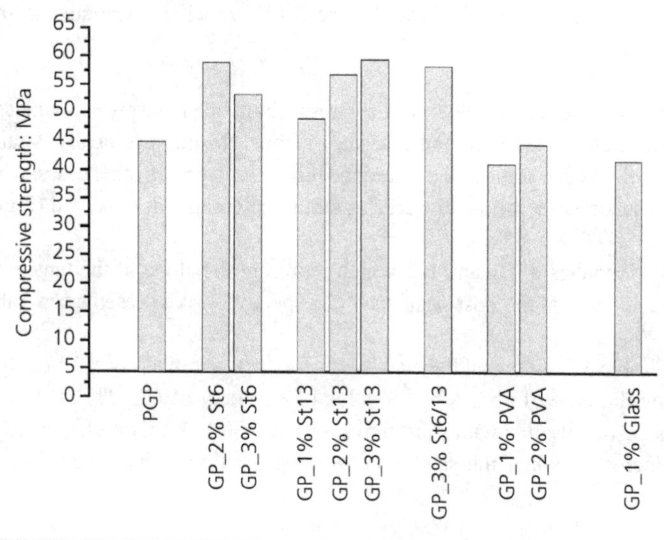

The results for the tensile strength (Figure 3.17) show the beneficial effect of the addition of steel fibres whereby the strength significantly increases as the fibre percentage is increased. The longer steel fibres (St13) are more efficient compared with the respective amount of

Figure 3.17 Tensile strength of geopolymer mixes with different types and amounts of fibres

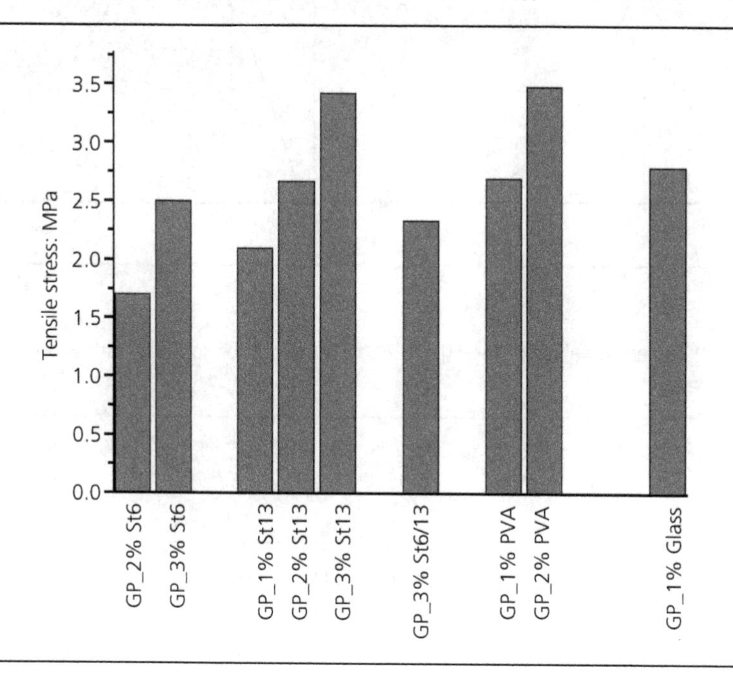

shorter steel fibres (St6), which is attributed to the enhancement of the bonding between the fibres and the geopolymer matrix. The PVA and the glass fibres were found to be even more efficient compared with the steel fibres (Figure 3.17), which is attributed to the increased aspect ratio of these types.

From the results of the direct tensile tests, the stress–strain behaviour was calculated and this is significantly affected by the type and amount of fibres. More specifically, strain-hardening behaviour was achieved in some of the examined mixes, whereas, in other cases, when the fibres were unable to successfully bridge the cracks, strain softening was observed (Figure 3.18).

The characteristic values of E_{el} and E_{cr}, which represent the slope of the tensile stress–strain curve at the elastic and at the post-peak stages, respectively, are presented in Table 3.5.

The results of Table 3.5 show that the elastic behaviour is significantly affected by the volume fraction and by the type of fibre. The modulus of elasticity of GP_2% St13 is almost 60% higher than the respective mix with short fibres (GP_2% St6). Also, the elastic modulus (E_{el}) is significantly increased when the steel fibre volume fraction is increased from 2% to 3%.

Figure 3.18 Typical tensile strain softening and strain hardening

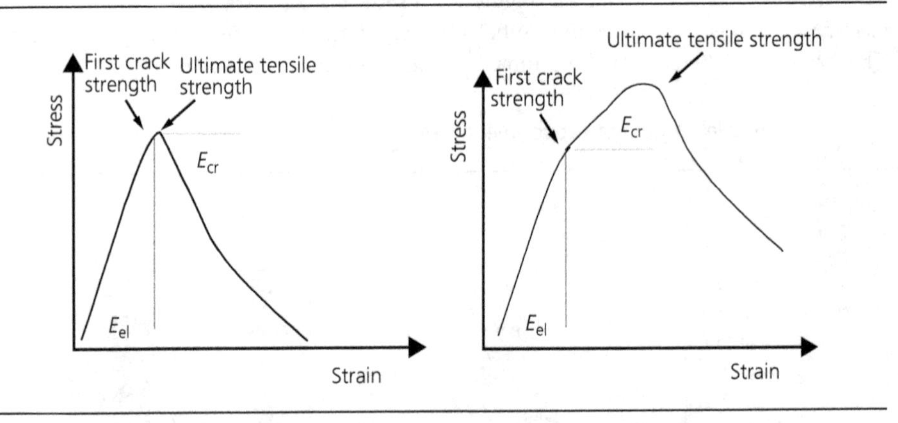

Table 3.5 Characteristic values of the E_{el} and E_{cr} for geopolymer concrete with various types of fibre reinforcement [31]

Modulus of elasticity	GP_2% St6	GP_3% St6	GP_1% St13	GP_2% St13	GP_3% St13	GP_3% St6/13	GP_1% PVA	GP_2% PVA	GP_1% glass
E_{el}: MPa	9855	19 978	9804	16 064	24 885	13 636	16 129	23 974	15 353
E_{cr}: MPa			431	304	317	665	235	4414	353

Similar observations can be made for the PVA fibres whereby the elastic modulus (E_{el}) quite significantly increases when the volume fraction of the fibres is increased from 1% to 2%.

The results for the post-cracking stage, show that for short (6 mm) fibres the performance is characterised by strain-softening behaviour, whereas by increasing the length of the fibres to 13 mm, strain hardening is achieved with increasing slope (E_{cr}) as the volume fraction is increased. The highest strain-hardening performance was achieved for the mix with 2% PVA fibres (GP_2% PVA).

3.3.4 Shear performance

The shear resistance of fibre-reinforced geopolymer beams with basalt fibre-reinforced polymer (BFRP) bars as longitudinal reinforcement was studied [33].

In this study, the shear resistance was studied in the absence of any shear links.

Extensive experimental work was conducted and analytical models were evaluated. The general conclusion is that the shear performance of geopolymer concrete beams is quite similar to the respective behaviour of conventional concrete beams.

Therefore, and in the absence of detailed studies in this field, it is recommended that analytical models presented in Section 2.3.3 for SFRC could also be used for geopolymer concrete until the development of new specific models for geopolymer concrete.

3.3.5 Punching shear of geopolymer concrete slabs

The punching shear performance has been recently studied and there have been experimental investigations using various mixes [34–37].

The main parameter examined is the amount of fibre reinforcement. It was found that, as the fibre reinforcement is increased, the punching shear capacity of the slabs is increased, as expected [34–35]. In another study [36], comparisons were made between geopolymer concrete and conventional concrete slabs reinforced with longitudinal bars (steel or fibre-reinforced polymer (FRP)) and higher punching performance was achieved in the case of geopolymer concrete, which was attributed to the higher mechanical characteristics of the geopolymer mix.

Geopolymer concrete slabs with nanosilica (NS) reinforced with steel fibres have also been examined under punching shear and the results show that significant enhancement of the punching shear strength is achieved with the use of NS and steel fibres [37]. This is attributed to the enhanced mechanical properties due to the improved bonding between the fibres and the geopolymer matrix.

There are no proposed models to date for the punching shear performance of geopolymer slabs. The punching shear is linked to the mechanical performance of the geopolymer mixes and, in the absence of a modified accurate model, the existing models for conventional concrete can be used.

3.3.6 Durability

The application of PVA fibre reinforced geopolymer concrete (PVAFRGC) for the repair and strengthening of RC beams was shown to be successful [23, 24, 38]. Beams reinforced with steel bars were examined (Figure 3.19).

The longitudinal reinforcement of the beams had 10 mm diameter with a yielding stress of 530 MPa. The reinforcement consisted of two deformed bars with a diameter of 10 mm (2Φ10) made of steel with a characteristic yielding stress value of 530 MPa. Two bars were placed on the tensile side and there were another two bars in the compressive zone with a gap near the middle of the span so as not to contribute to compressive reinforcement. Shear links of 8 mm diameter and yield strength 350 MPa were placed in the shear span with a spacing of 90 mm. The compressive strength of the conventional concrete of the beam was 32 MPa.

Numerous specimens were examined and, in some of the examined specimens, casting stopped at 150 mm and at a depth of 175 mm and PVAFRGC was used for the casting of the remaining 50 mm and 25 mm layers, respectively, to represent a case of concrete repair.

The same mix design as that presented in Table 3.3 was used for the PVAFRGC. The mechanical characteristics of the material were evaluated by means of standard compressive tests: flexural

Figure 3.19 RC beams with PVAFRGC layers [23, 38]

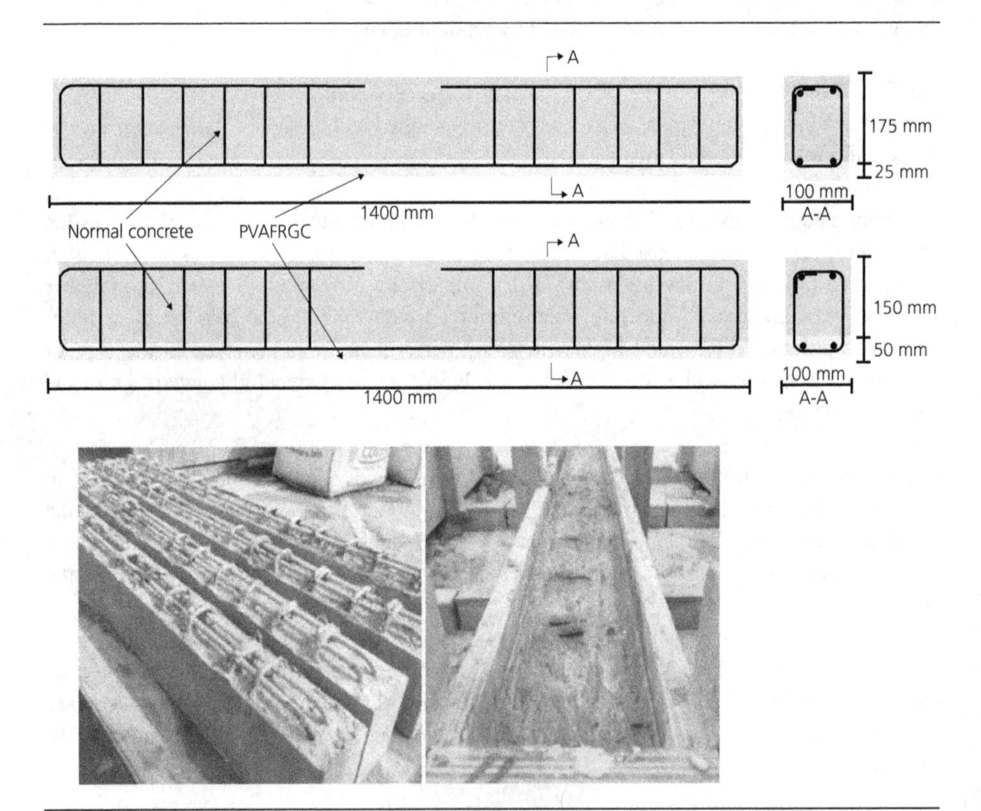

and direct tensile tests were conducted. The compressive strength (on the day of the testing of the beams) was found to be 45 MPa and the tensile strength was found to be 3 MPa.

An accelerated corrosion test was carried out to simulate the effect of corrosion (Figure 3.20). Some of the examined specimens were exposed to accelerated corrosion while other specimens were left uncorroded to act as the control specimens. For the accelerated corrosion method, the beams were immersed in a 5% sodium chloride solution which was placed up to a height of 5 cm (Figure 3.20). A constant current of 300 mA was applied for 90 days between the reinforcement bar and a copper mesh connected to the negative terminal of the DC power supply (Figure 3.20).

At the end of this procedure, the specimens were tested with the application of a load to the middle of the span. RC beams without PVAFRGC without corrosion (Figure 3.21(a)) and after the application of corrosion (Figure 3.21(b)) were examined [23, 38].

Figure 3.20 Experimental set-up used for the accelerated corrosion [23, 38]

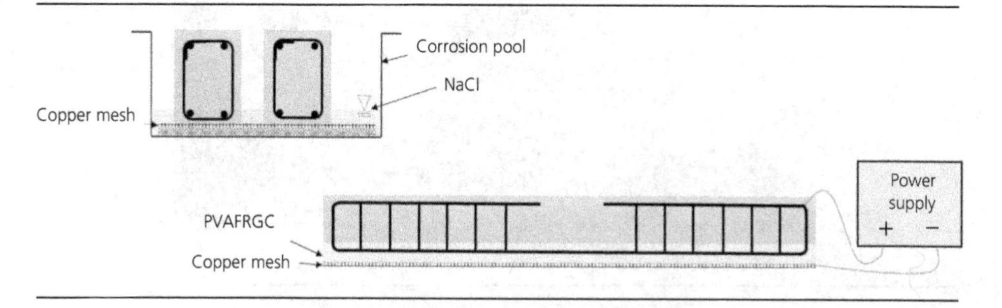

Figure 3.21 Conventional RC beams tested (a) without corrosion and (b) after the effect of corrosion [23, 38]

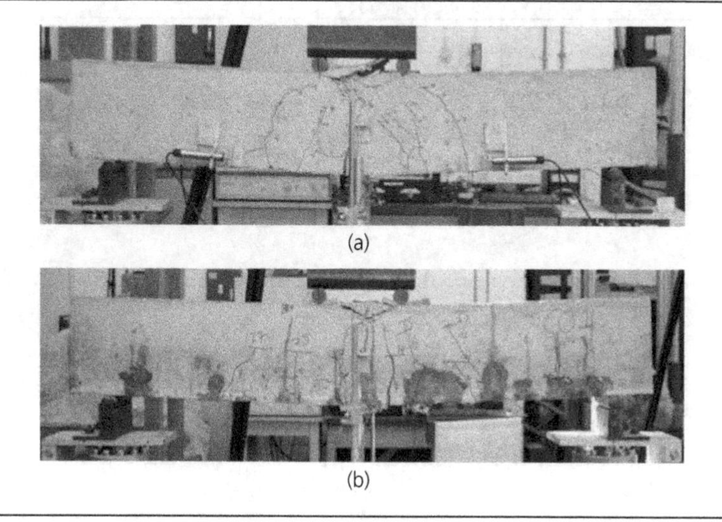

Respective specimens with 25 mm and 50 mm PVAFRGC layers without corrosion and after the application of corrosion were also examined (Figures 3.22 and 3.23) [23, 38].

The results for load against mid-span deflection are presented in Figures 3.24–3.26 [23, 38].

Figure 3.22 RC beams with 25 mm PVAFRGC tested (a) without corrosion and (b) after the effect of corrosion [23, 38]

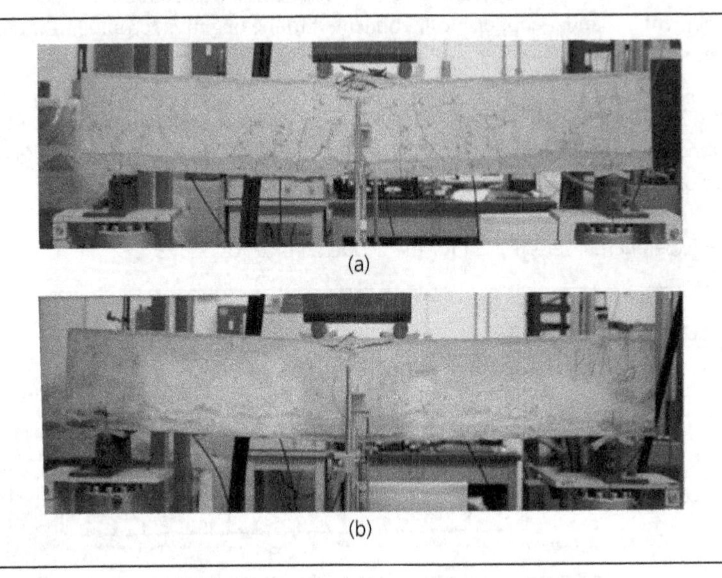

(a)

(b)

Figure 3.23 RC beams with 50 mm PVAFRGC tested (a) without corrosion and (b) after the effect of corrosion [23, 38]

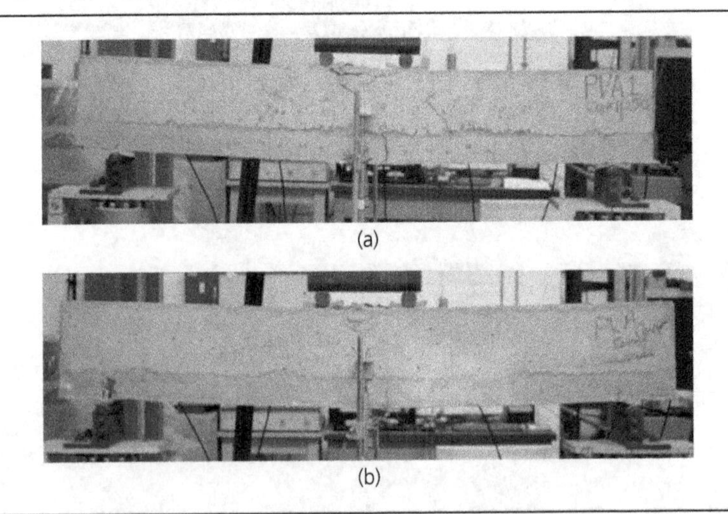

(a)

(b)

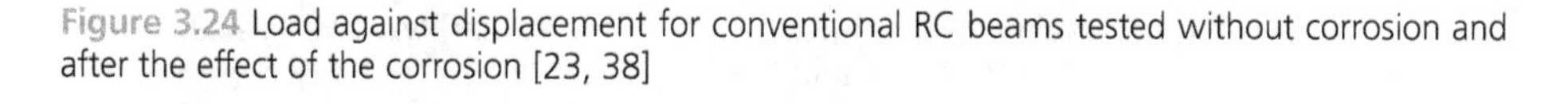

Figure 3.24 Load against displacement for conventional RC beams tested without corrosion and after the effect of the corrosion [23, 38]

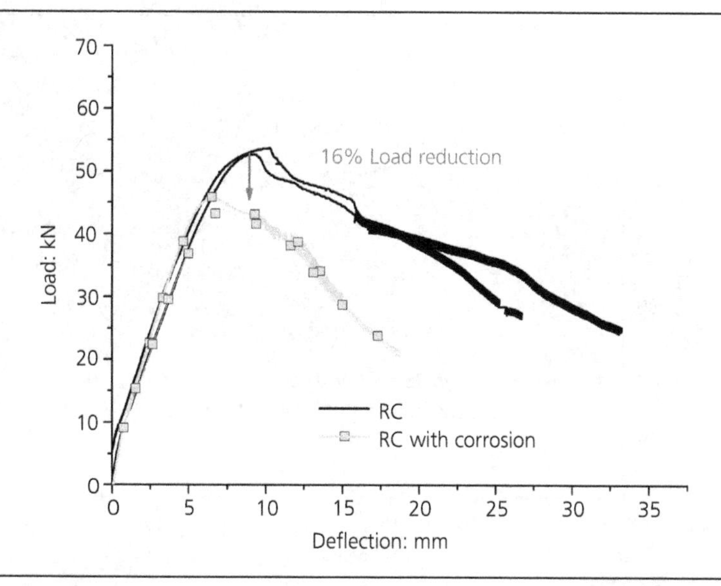

The results of Figure 3.24 show that the ultimate flexural load of the conventional RC beams is reduced by an average of 16% when the accelerated corrosion is applied.

The reduction of the ultimate load capacity of the beams due to the effect of the corrosion was controlled by the use of 25 mm repair PVAFRGC since the reduction of the ultimate load capacity was reduced to 11% (Figure 3.25). This was further reduced to 8% (Figure 3.26) when a 50 mm PVAFRGC layer was used as a repair material.

In addition, all the examined beams with PVAFRGC show improved structural performance compared with the respective results of conventional RC beams (Figures 3.24–3.26) and this is attributed to the enhanced tensile stress–strain characteristics of the PVAFRGC.

The beneficial effect of the presence of the PVAFRGC for the protection of the steel bars from corrosion was further confirmed by measuring the mass loss of the steel bars after the induced corrosion method. More specifically, at the end of the testing of the RC beams, samples of the steel bars were removed and they were used to evaluate the effect of the corrosion. Mechanical chemical and electrolytic techniques were used to remove the corrosion products and the mass loss was calculated by subtracting the mass of the corroded elements from the mass of the uncorroded ones [23, 38]. The average results for the examined specimens are presented in Figure 3.27.

The results show that the effect of the corrosion is linearly reduced from 8.5% to 7% and then to 5.5% when part of the RC section is replaced by 25 mm and 50 mm PVAFRGC layers, respectively.

Figure 3.25 Load against displacement for RC beams with a 25 mm PVA layer tested without corrosion and after the effect of the corrosion [23, 38]

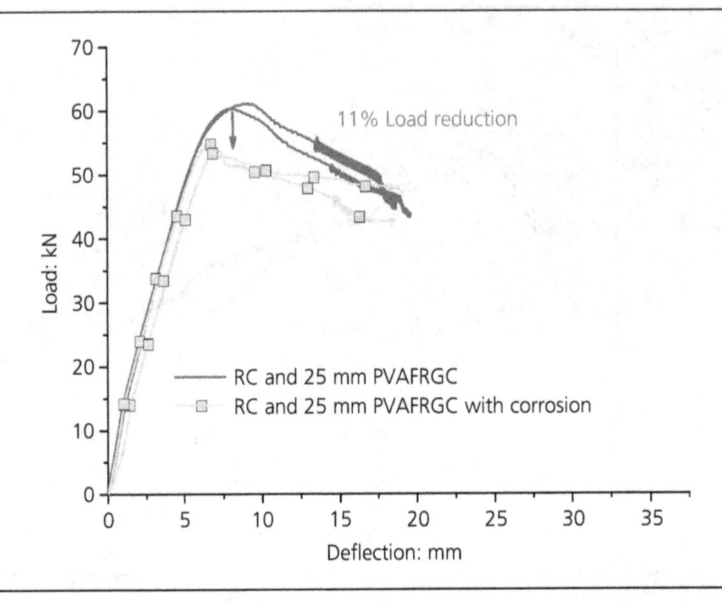

Figure 3.26 Load against displacement for RC beams with a 50 mm PVA layer tested without corrosion and after the effect of the corrosion [23, 38]

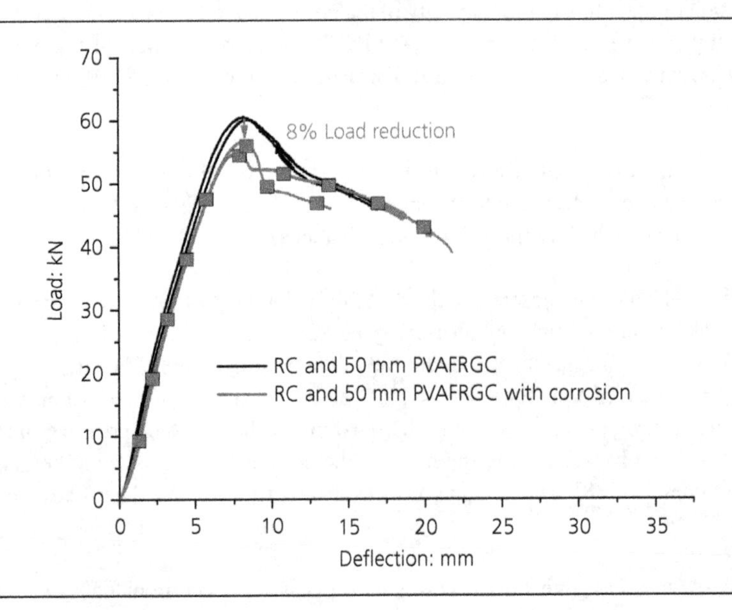

Figure 3.27 Mass loss reduction in the steel bars due to the effect of corrosion

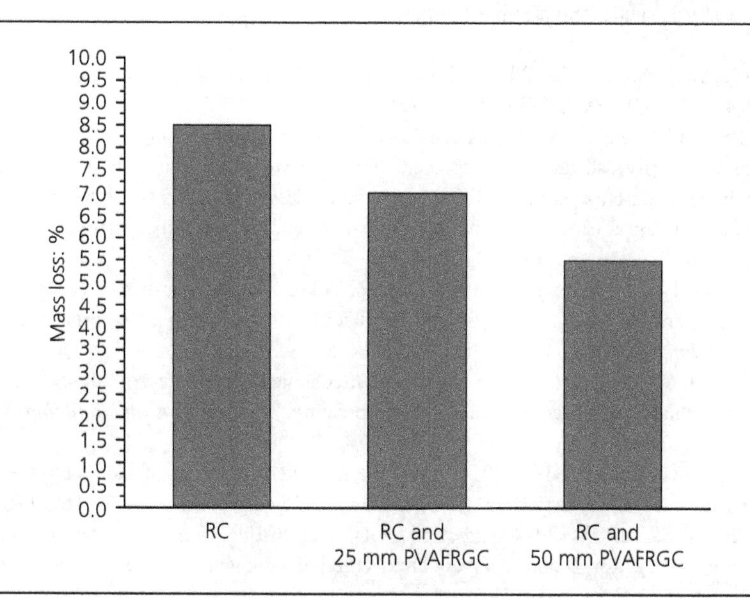

REFERENCES

1. Kühl H (1930) *Zementchemie*. Verlag Technik, Berlin, Germany. Band III; 1958 or Zement 19.
2. Chassevent L (1937) 17. Congress of Industrial Chemistry. 147 (Paris).
3. Purdon A (1940) The action of alkalis on blast furnace slag. *Journal of the Society of Chemical Industry* **59**: 191–202.
4. Davidovits J (1988) Ancient and modern concretes: what is the real difference? *Concrete International* **9**: 23–29.
5. Krivenko P (1986) Synthesis of cementitious materials of the Me_2O-MeO-Me_2O_3-SiO_2-H_2O system with required properties. DSc(Eng) thesis, *KISI Publis*. Kiev, Ukraine.
6. Palomo A, Grutzeck MW and Blanco MT (1999) Alkali-activated fly ashes – a cement for the future. *Cement and Concrete Research* **29(8)**: 1323–1329.
7. Mostafa NY, El-Hemaly SAS, Al-Wakeel EI, El-Korashy SA and Brown PW (2001) Characterization and evaluation of the hydraulic activity of water-cooled slag and air-cooled slag. *Cement and Concrete Research* **31(6)**: 899–904.
8. Li D, Xu Z, Luo Z, Pan Z and Lin C (2002) The activation and hydration of glassy cementitious materials. *Cement and Concrete Research* **32(7)**: 1145–1152.
9. Pal SC, Mukherjee A and Pathak SR (2003) Investigation of hydraulic activity of ground granulated blast furnace slag in concrete. *Cement and Concrete Research* **33(9)**: 1481–1486.
10. Chao Li, Henghu Sun and Longtu Li (2010) A review: the comparison between alkali-activated slag (Si+Ca) and metakaolin (Si+Al) cements. *Cement and Concrete Research* **40(9)**: 1341–1349.
11. Bernal SA, Mejía de Gutierrez R, Ruíz F, Quiñones H and Provis JL (2012) High-temperature performance of mortars and concretes based on alkali-activated slag/metakaolin blends. *Materiales de Construccion* **62(308)**: 471–488.

12. Mejía JM, Mejía de Gutierrez R and Puertas F (2013) Rice husk ash as a source of silica in alkali-activated fly ash and granulated blast furnace slag systems. *Materiales de Construccion* **63(311)**: 361–375.
13. Provis J, van Deventer JSJ (2014) *Alkali Activated Materials. State of the Art Report. RILEM TC 224-AAM*. Springer, Dordrecht, the Netherlands.
14. Deb PS, Nath P and Sarker PK (2014) The effects of ground granulated blast-furnace slag blending with fly ash and activator content on the workability and strength properties of geopolymer concrete cured at ambient temperature. *Materials and Design* **62**: 32–39.
15. Chi M and Huang R (2013) Binding mechanism and properties of alkali-activated fly ash/slag mortars. *Construction and Building Materials* **40**: 291–298.
16. Marjanović N, Komljenović M, Baščarević Z, Nikolić V and Petrović R (2015) Physical–mechanical and microstructural properties of alkali-activated fly ash–blast furnace slag blends. *Ceramics International: Part B* **41(1)**: 1421–1435.
17. Lee NK and Lee HK (2013) Setting and mechanical properties of alkali-activated fly ash/slag concrete manufactured at room temperature. *Construction and Building Materials* **47**: 1201–1209.
18. Jang JG, Lee NK and Lee HK (2014) Fresh and hardened properties of alkali-activated fly ash/slag pastes with superplasticizers. *Construction and Building Materials* **50**: 169–176.
19. Nath P and Sarker PK (2014) Effect of GGBFS on setting, workability and early strength properties of fly ash geopolymer concrete cured in ambient condition. *Construction and Building Materials* **66**: 163–171.
20. Al-Majidi MH, Lampropoulos A, Cundy A and Meikle S (2016) Development of geopolymer mortar under ambient temperature for in situ applications. *Construction and Building Materials* **120**: 198–211.
21. Al-Majidi MH, Lampropoulos A and Cundy A (2017) Tensile properties of a novel fibre reinforced geopolymer composite with enhanced strain hardening characteristics. *Composite Structures* **168**: 402–427.
22. Al-Majidi MH, Lampropoulos A and Cundy A (2017) Steel fibre reinforced geopolymer concrete (SFRGC) with improved microstructure and enhanced fibre-matrix interfacial properties. *Construction and Building Materials* **139**: 286–307.
23. Al-Majidi MH, Lampropoulos A, Cundy A, Tsioulou O and Alrekabi S (2018) A novel corrosion resistant repair technique for existing reinforced concrete (RC) elements using polyvinyl alcohol fibre reinforced geopolymer concrete (PVAFRGC). *Construction and Building Materials* **164**: 603–619.
24. Al-Majidi MH, Lampropoulos A, Cundy A, Tsioulou O and Alrekabi S (2019) Flexural performance of reinforced concrete beams strengthened with fibre reinforced geopolymer concrete under accelerated corrosion. *Structures* **19**: 394–410.
25. Nguyen T (2017) *Durability of Reinforced GGBS/FA Geopolymer Concretes*. PhD thesis, Loughborough University, UK.
26. BS EN 450-1 (2012) Fly ash for Concrete: Definition, Specifications and Conformity Criteria. BSI, London, UK.
27. Rafeet A, Vinai R, Soutsos M and Sha W (2017) Guidelines for mix proportioning of fly ash/GGBS based alkali activated concretes. *Construction and Building Materials* **147**: 130–142.
28. Pavithra P, Srinivasula Reddy M, Dinakar P, Hanumantha Rao B, Satpathy BK and Mohanty AN (2016) A mix design procedure for geopolymer concrete with fly ash. *Journal of Cleaner Production* **133**: 117–125.
29. DIN 1045 (1988) Beton und Stahlbeton. Beton Verlag GMBH, Koln.

30. Mermerdas K, Manguri S, Eddin Nassani D and Mahdi Oleiwi S (2017) Effect of aggregate properties on the mechanical and absorption characteristics of geopolymer mortar. *Engineering Science and Technology, an International Journal* **20**: 1642–1652.
31. Al-Majidi MH, Lampropoulos A and Cundy AB (2017) Tensile properties of a novel fibre reinforced geopolymer composite with enhanced strain hardening characteristics. *Composite Structures* **168**: 402–427.
32. Al-Majidi MH, Lampropoulos A and Cundy AB (2017) Steel fibre reinforced geopolymer concrete (SFRGC) with improved microstructure and enhanced fibre-matrix interfacial properties. *Construction and Building Materials* **13**: 286–307.
33. Tran TT, Pham TM and Hao H (2020) Effect of hybrid fibers on shear behaviour of geopolymer concrete beams reinforced by basalt fiber reinforced polymer (BFRP) bars without stirrups. *Composite Structures* **243**: 112236.
34. Karunanithi S (2017) Experimental studies on punching shear and impact resistance of steel fibre reinforced slag based geopolymer concrete. *Advances in Civil Engineering* **1**: 1–9.
35. Sadawy MA, Faried AS and El-Ghazaly HA (2021) Investigating the impact of punching shear strength on geo-polymer concrete slabs with openings. *Design Engineering* **(8)**: 9148–9161.
36. Mohmmad SH, Güls ME and Çevik A (2022) Punching shear behaviour of geopolymer concrete two-way slabs reinforced by FRP bars under monotonic and cyclic loadings. *Advances in Structural Engineering* **25(3)**: 453–472.
37. Eren NA (2022) Punching shear behavior of geopolymer concrete two-way flat slabs incorporating a combination of nano silica and steel fibres. *Construction and Building Materials* **346**: 128351.
38. Al-Majidi MH (2017) *Development of Fibre Reinforced Geopolymer Concrete (FRGC) Cured Under Ambient Temperature for Strengthening and Repair of Existing Structures*. PhD thesis, University of Brighton, UK.

Lampropoulos A
ISBN 978-0-7277-6556-7
https://doi.org/10.1680/fchs.65567.075

Chapter 4
Critical evaluation and main considerations for the selection of structural concrete types and selected case studies

4.1. Mechanical performance

Mechanical performance is one of the key parameters for the selection of the appropriate materials for the design and construction of new structures.

The addition of fibre reinforcement in FRC can be used to enhance the tensile stress characteristics, in particular the post-cracking residual tensile stresses, because of the bridging effect offered by the fibres. When appropriate types of fibres are used at high volume fractions (normally 2% and above) in addition to the appropriate cementitious matrix, the ultimate tensile stress can also be quite significantly enhanced and strain-hardening characteristics can be provided. The compressive stress characteristics may also be enhanced but this is not the main reason for the use of fibre reinforcement in concrete.

For UHPFRC, the same principles as with FRC are followed but, in this case, specialised materials are required for the cementitious matrix in addition to a high volume fraction of fibres (normally steel fibres in the range of 2%–6%). For the cementitious matrix, high-strength cement in addition to fine materials are required for the 'particle packing' and for the development of the ultra-high-strength cementitious matrix. The significantly enhanced strength and density of the cementitious matrix contributes to the improvement of the fibre-to-matrix bond, which leads to significantly enhanced energy absorption and superior characteristics in both compression and tension. The typical compressive strength of UHPFRC is in the range of 150–200 MPa and the tensile strength can be higher than 10 MPa, and strain-hardening characteristics and significantly enhanced post-crack residual strength are normally achieved.

In the case of fibre-reinforced geopolymer concretes (FRGC), the mechanical characteristics are significantly affected by the composition of the mix, the selection of the raw materials and the curing conditions. The selection of the appropriate materials can lead to high-strength mixes with compressive strength above 50 MPa, while heat curing can lead to enhancement of the mechanical characteristics and to significant reduction of the required curing time. The use of fibre reinforcement can lead to enhancement of the tensile stress characteristics and, in particular, the energy absorption and the residual post-cracking stresses. The geometry of the fibres is an important parameter and increased aspect ratio was found to be beneficial for the fibre-to-geopolymer concrete bond and, subsequently, for the tensile stress characteristics. Different types of fibres were used; PVA, glass and steel fibres were found to be the most efficient.

The general conclusion is that FRGC can have high strength characteristics but the superior properties of UHPFRC cannot be achieved with the current technology, and the selection of the appropriate type of material is highly dependent on the requirements and on the expected loading conditions. The use of UHPFRC could be ideal for thin elements under extreme loading conditions and could be combined with the use of additional steel bars if improved flexural performance and ductility is required.

4.2. Durability

The durability performance of FRC with different types of fibres was studied under various deterioration processes such as chlorides, carbonation, alkali–silica reaction, high temperatures, and freeze and thaw tests. The main conclusion is that the durability is enhanced, mostly owing to the control of the cracks offered by the crack-bridging of the fibre, which leads to reduced entry of substances (for example, water, chlorides and carbon dioxide) and limited deterioration.

In the case of UHPFRC, the durability is further enhanced by the low water to binder ratio of the mix and the presence of fine aggregates. The corrosion resistance of UHPFRC has also been found to be significantly enhanced compared with conventional concrete. Accelerated carbonation tests have shown that UHPFRC exhibits minor or no signs of carbonation depending on the mix. Chloride migration of UHPFRC mixes has also been found to be significantly lower compared with conventional concrete owing to the high cement content which leads to the presence of an increased amount of tricalcium aluminate.

The use of fibre-reinforced geopolymer concrete has also been found to offer significantly enhanced protection of the steel reinforcement under accelerated corrosion tests. This is attributed to the dense matrix and the composition of the examined fibre-reinforced geopolymer mixes.

4.3. Environmental and economic considerations

The environmental and economic considerations are two of the key drivers for the selection of the appropriate materials that are decided at the 'conceptual design' stage. The environmental impact requires the consideration of the effect of the examined materials during the various stages of the process, that is, production, construction, usage and end of life, and a life cycle analysis is normally conducted to accurately evaluate this.

The use of FRC, UHPFRC and FRGC have significant benefits in terms of mechanical performance and durability. However, the cost and the environmental impact of the production and construction of these materials is significantly affected by the materials that are used and by the construction process. The cost of the materials is markedly increased by the addition of steel with a high percentage of steel fibres. For example, in case of UHPFRC, for which a volume fraction of 2%–6% (157–471 kg/m^3) is normally used, the average cost of the fibres is in the range of 3–4 £/kg, so the cost of the fibres for only 1 m^3 of UHPFRC is around £470–1880. This cost seems very high when compared with conventional concrete. However, if additional parameters are considered such as the reduction in the required cross-section of the structural elements, the reduction in the labour costs, the reduced need for maintenance and the longer life expectancy, this can be a cost effective and preferred option for specific applications. The same applies to the environmental impact of the use of UHPFRC as it requires a very high amount of cement. However, a reduction in the cross-section of the structural elements can be achieved, which leads to a reduced demand for construction materials.

The use of geopolymers and fibre-reinforced geopolymer concretes have significant environmental benefits owing to the replacement of cement by waste materials, which leads to significant reduction in the carbon dioxide emissions linked to cement production, while, at the same time, significantly reducing pollution issues linked to the disposal of waste materials in landfill sites. The cost of geopolymer concrete is normally not dissimilar to the cost of conventional concrete but this depends on the availability of raw materials in each country and the type of activators that are used.

In terms of the production process, FRC, UHPFRC and FRGC can be used either for cast in situ applications or for precast elements. The use of precast elements can, in some cases, be the preferred solution because of the enhanced speed of production (especially in case of heat curing), enhanced quality assurance and the reduction in material wastage, but a detailed analysis and the consideration of various parameters is normally required (for example, a life cycle analysis) to select the most appropriate process and material.

4.4. Characteristic case studies

4.4.1 SFRC case studies

In this section, two applications are presented for the use of SFRC in construction. The first is focused on a pilot study on the use of sustainable SFRC for the construction of pavements known as Ecolanes [1], while the second is for a tunnel.

4.4.1.1 Sustainable SFRC pavements

Introduction and project description The case studies which will be presented in this section were conducted as part of the research project 'Ecolanes', which was funded by the 6th EU Research Framework Programme (FP6). The key partners of this project included four universities (The University of Sheffield (UK), Akdeniz University (Turkey), Technical University 'Gheorghe Asachi' Iasi (Romania), Cyprus University of Technology (Cyprus)) and several industrial partners, associations and end-users (European Tyre Recycling Association (France), Aggregate Industries (UK), Ltd (United Kingdom), Antalya Municipality (Turkey), Compania Nationala de Autostrazi si Drumuri Nationale din Romania (Romania), Adriatica Riciclaggio e Ambiente s.r.l (Italy), Public Works Department, Ministry of Communications and Works (Cyprus), Scott Wilson Ltd (UK)).

The main aim of this project was to apply roller-compaction techniques and recycled materials for the construction of sustainable and low-cost concrete pavements. The use of recycled tyre steel cord (RTC) fibres as concrete reinforcement was extensively studied under this project and suitable materials were developed for this application.

During the first period of the project, the fibres were processed so as to make them suitable for their use in concrete. Cleaning processes were conducted and the fibres were classified based on their length distribution. Then appropriate mixes were developed using suitable volume fractions, and flexural tests were conducted to characterise the material properties. Finally, an appropriate construction process was developed and demonstration pavements were constructed in the UK, Romania, Turkey and Cyprus.

Material properties Recycled steel fibres from post-consumer tyres were used in this project. Unprocessed fibres (Figure 4.1(a)) were cleaned and processed to the desired lengths, as specified for the purpose of this project (Figure 4.1(b)).

Mixes with crushed and recycled aggregates were examined. The recycled aggregates were produced from previous reclaimed crushed concrete. The crushed aggregates offer the required properties for the stability of the roller-compacted concrete. Also, conventional steel fibres and recycled tyre steel cord fibres were used to evaluate the effect of the fibres on the mechanical characteristics of the mixes [2]. The conventional steel fibres had a length of 54 mm and a

Figure 4.1 (a) Unprocessed and (b) processed recycled steel fibres and (c, d) indicative flexural test results of the examined mixes [1]

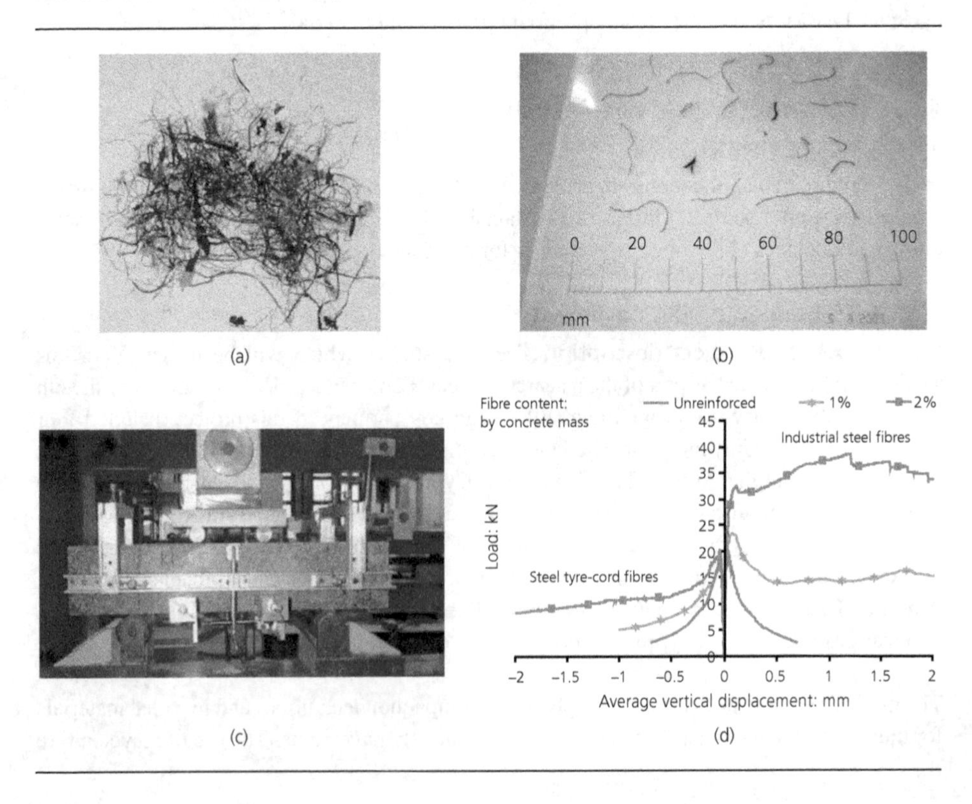

diameter of 1 mm, whereas the RTC had a diameter of 0.13 mm. Two different types (Ecolanes Class B and Ecolanes Class C) based on their length distribution were examined. Ecolanes Class B fibres consisted of at least 40% of fibres of length 15–25 mm and less than 20% with length more than 25 mm. Ecolanes Class C fibres consisted of at least 80% of fibres of length 1–15 mm length and less than 10% with length more than 25 mm. A detailed description of the examined mix compositions is presented in Table 4.1.

The results of the flexural testing of the examined mixes show that the roller-compacted concrete (RCC) with conventional steel fibres has improved flexural performance compared with the respective specimens. An increment in the amount of RTC fibres was found to be beneficial and the behaviour of the specimens with high amounts of RTC fibres (mixes D and F) was similar to the results for the specimen with a lower amount of conventional fibres (mix B). The addition of recycled aggregates did not significantly affect the results and the use of longer RTC fibres (Ecolanes Class B fibres) was found to be more efficient for the post-cracking behaviour of the examined material compared with the shorter fibres (Ecolanes Class C fibres).

Construction process A roller-compaction method was used for the construction process. The dry mix was placed in a truck (Figure 4.2(a)) and it was transferred to the construction place. Then the dry mix was placed on the pavement using a paver (Figure 4.2(b)). Joints were introduced (Figure 4.3(a)) to prevent cracking of the pavements which were filled with emulsion (Figure 4.3(b)). The final compaction was conducted using a heavy-duty dual drum vibrating roller which passed over the RCC pavement several times (Figure 4.4) [1].

Four demonstration pavements were constructed, in London (UK), Gura Humorului (Romania), Antalya (Turkey) and Pafos (Cyprus).

In London, the pavement was constructed in April 2009 and was subjected to controlled heavy-goods traffic. It had a surface area of 300 m^2 and comprised three layers. The foundation was constructed using cement-bound granular materials and had a thickness of 150 mm. The second layer was constructed using SFRC with 5% (by mass) RTC fibres and had a thickness

Table 4.1 Examined concrete mixes [1]

	Aggregates		Cement	Fibres	
Mix ID	Type	Amount: kg/m^3		Type	Amount: kg/m^3
A	Conventional	2092	300		
B	Conventional	2081		Conventional	50
C	Conventional	2064		Class C RTC	75
D	Conventional	2036		Class C RTC	150
E	Recycled	1893		–	–
F	Recycled	1870		Class B RTC	150

Figure 4.2 (a) RCC loaded to the trucks and (b) paver for laying [1, 2]

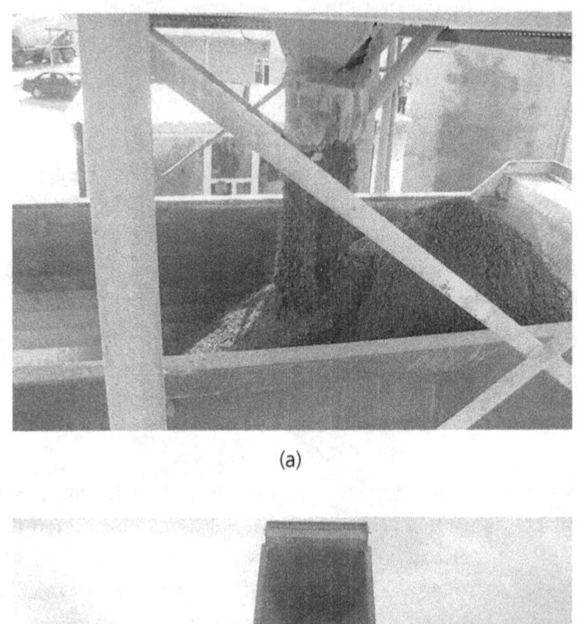

(a)

(b)

of 170 mm. At the top of the pavement, an asphalt overlay was constructed with 70 mm thickness [1].

The second pilot pavement, which was constructed in May 2019 in Gura Humorului (Romania), was focused on the rehabilitation of an existing heavily used national road. It consisted of three layers: for the foundation layer, which had thickness of 300 mm, basalt material was used; for the second layer, concrete was used with three different thicknesses (180 mm, 230 mm and 280 mm) at various sections; and the third layer was the asphalt overlay. In the second layer, two lanes were constructed, one using plain RCC and one using dry SFRC with 3% RTC fibres, with a width of 9.5 m and a total length of 150 m [1].

Figure 4.3 (a) Joints to prevent cracking and (b) filling with emulsion [1]

(a)

(b)

The third demonstration was conducted in Antalya (Turkey) (Figure 4.5) and comprised full rehabilitation of an existing urban road. The pavement consisted of four layers. The foundation layer was made of ballast and had a thickness of 200 mm. The base course was made of broken stone and had a thickness of 100 mm. The third layer was a concrete base with a thickness of 190 mm, and the fourth layer, which was constructed on the top, was the asphalt overlay with 40 mm thickness. The concrete base was constructed in four different joint-less sections. The first section was 70 m long and 5.1 m wide and was made of SFRC with 3% (by mass) RTC fibres; the second section was 40 m long and 5.1 m wide and was made of SFRC with 2% (by mass) commercial steel fibres; the third section was 150 m long and 3.5 m wide and was made of plain dry mix concrete; and the fourth section was 150 m long and 3.5 m wide and was made of SFRC with 3% (by mass) RTC fibres [1].

Figure 4.4 Roller compaction of the pavements [1]

Figure 4.5 Construction of pavement in Antalya (Turkey)

(a)

(b)

The fourth pavement was constructed in Pafos (Cyprus) and was part of the rehabilitation of an existing road in a rural area (Figure 4.6). The pavement had a length of 40 m and a width of 6 m and consisted of two layers which were placed on the top of the existing pavement to give a total thickness of 100 mm. The first of the two layers, which was constructed in contact with the existing pavement (after a slight milling of the surface), was made of SFRC with 2% (by mass) RTC fibres, and asphalt was placed on the top [1].

Concluding remarks These demonstration projects focused on the use of RTC fibres from post-consumer tyres as a replacement for conventional steel fibres in the construction of pavements. Dry mixes were developed and the method of RCC was applied to reduce construction costs and to offer enhanced properties.

Figure 4.6 Construction of pavement in Pafos (Cyprus)

(a)

(b)

The examined construction process offers cost and energy reduction and reduced maintenance requirements, so is a very attractive alternative sustainable method for the construction of pavements [1, 2].

More specifically, the proposed methodology is aimed at reducing the cost of construction by 10–20%, reducing construction time by approximately 15% and reducing energy consumption by around 40%, in addition to requiring less maintenance and having additional sustainability benefits owing to the use of waste materials [1, 2]. Potential difficulties and issues are mostly linked to the mixing of the RTC fibres and the efficiency of using this type of fibre compared with conventional steel fibres. A previous study [3] has highlighted potential issues caused by fibre balling, which is mostly linked to the irregular shape of the RTC fibres owing to their length distributions. A recent study in this field [4] has shown that cleaning the RTC fibres and reducing the number of short-length fibres can significantly improve the concrete quality and so offer enhanced mechanical characteristics [4].

4.4.1.2 The Lee Tunnel

Introduction and project description The Lee Tunnel was opened in 2016 and it runs for 4.3 miles. It is one of the two tunnels which will collectively capture an average of 39 megatonnes of sewage a year. This serves one of the most polluting combined sewer overflows in London, UK; it is part of a sewerage network which has been in place for more than 150 years [5].

The construction of the Lee Tunnel, at a cost of £711 million, was designed to tackle discharges from London's largest combined sewer overflows at Abbey Mills Pumping Station in Stratford, which accounts for 40% of the total discharge.

The Lee Tunnel is the deepest tunnel in London. It was constructed under very high underground pressures passing through 6.9 km of abrasive ground, without any other shafts along the way [6].

Material properties The material used for the casting of the tunnel lining was a C50 concrete reinforced with 40 kg/m^3, 60 mm long double-hooked fibres.

The design of the tunnel segments required calculation of the properties of the materials. Flexural tests of standard prisms are normally required to determine characteristic values of the flexural strength of SFRC (Section 1.4.1), and standard cubes or cylinders are tested to determine the compressive characteristics.

For the design in compression, use of the simplified stress–strain model of Figure 4.7 is proposed by TR63 [6, 7]. The use of the characteristic compressive strength value (f_{ck}) is proposed for the design for SLS, whereas, for the ULS, the use of the design strength value is proposed (f_{cd}). These models are applicable to strength classes up to C50/60.

For the tensile and flexural strength characteristics, the results of the flexural testing of the beams are used and f_{r1} is calculated (Section 1.4.1).

Figure 4.7 Proposed stress–strain model for the design of concrete in compression [6, 7]

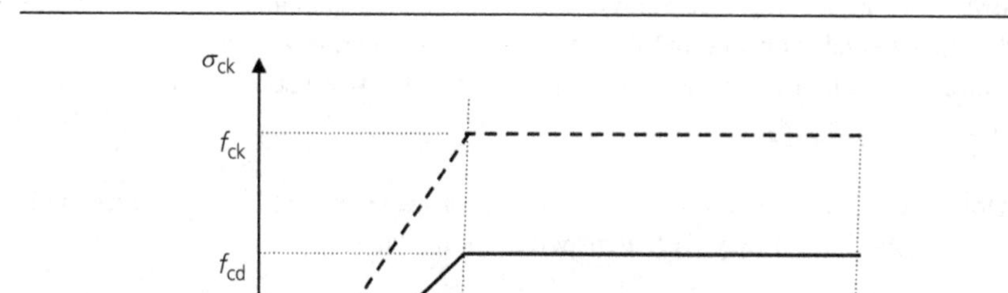

For the design at SLS, the tensile strength (f_t) is taken to be 0.45 f_{r1}. The exact stress–strain model used for the design depends on the crack width. For crack widths up to 0.5 mm, a constant stress–strain graph in tension is considered, and therefore the stress at the bottom of the tensile side (f_b) is taken to be $f_t = 0.45f_{r1}$. For SLS, crack opening (w_s) is considered up to 0.2 mm, whereas for ULS, the ultimate crack opening (w_u) depends on the type of fibre. According to Model Code 2010, the stress at this stage (f_{bu}) at the ULS depends on the value f_{r3} (derived from the prism tests) and is given by Equation 4.1 [6].

$$f_{bu} = 0.5 \times f_{r3} - 0.2 \times f_{R1} \tag{4.1}$$

The strain distribution together with the respective elastic and plastic stress distributions are illustrated in Figure 4.8.

For the design of the tunnel presented in this section, the values of f_{r1} = 3.45 MPa, f_{r3} = 3.00 MPa and w_u = 2.5 mm were used.

Details about the methods of analysis that were used for the design of the SFRC lining are available in a published study [6].

Figure 4.8 Distribution of (a) strains, (b) elastic and (c) plastic stresses along the cross-sectional depth [6]

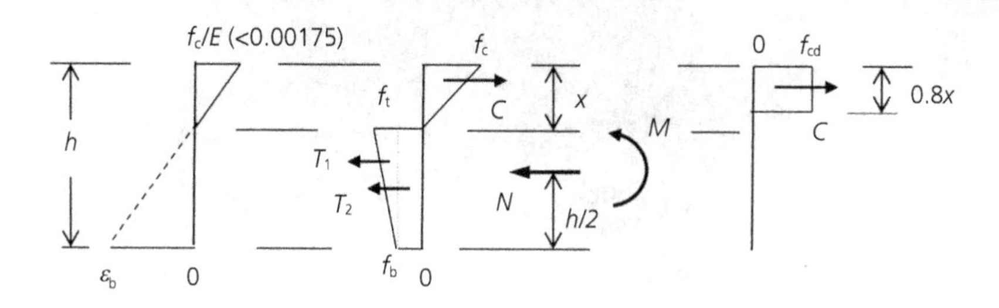

Construction process A tunnel boring machine (TBM) was used for the construction process. The TBM was used for the excavation of the bore which has a diameter just under 9 m and led to the final tunnel with internal diameter 7.2 m. A waterproofed primary lining of 350 mm thickness was initially placed for waterproofing and it was supplemented by a secondary 300 mm thick lining [8].

The 350 mm thick SFRC primary lining segments for the project were produced complete with cast-in elastomeric gaskets and they were transported to the site [9].

When the TBM was less than 30 m away from the shaft, the material and the primary lining segments were transported to the working area and they were fixed as the TBM was progressing [9].

Instead of using traditional reinforcement and sprayed concrete, the Lee Tunnel's secondary lining uses a patented FRC, placed using a travelling formwork in 30 m sections (Figure 4.9).

Concluding remarks The Lee Tunnel is the largest diameter and deepest tunnel in London and its novel shaft design led to the saving of 1500 tonnes of steel, reduced construction time and reduced risk of damage to the shaft linings caused by steel corrosion over the 80+ year working life of the sewer (MottMacDonald). The five large-diameter shafts were designed to meet watertightness requirements and were constructed to very exacting vertical tolerances through the use of bespoke concrete mixes for both the primary and inner lining of the walls, specialist diaphragm walling equipment and the construction of post-compressed fibre-reinforced concrete inner linings with double-sided slip-form shutters [9].

After the completion of the initial TBM-driven lining construction, a SFRC secondary lining was cast in situ to offer enhanced water resistance and an improved smooth and durable surface aimed at providing optimum hydraulic performance [9].

Figure 4.9 Construction of the Lee Tunnel [6]

4.4.2 UHPFRC case studies

4.4.2.1 The UHPFRC Footbridge in Lužec nad Vltavou

Introduction and project description This section is focused on the construction of a bridge which was built in 2015 to connect the towns of Lužec nad Vltavou and Bukol in the Czech Republic. The bridge was designed by a team of engineers and architects led by the architect and bridge engineer Petr Tej, the architect Marek Blank and the bridge engineer Jan Mourek [10]. The bridge, which runs over a river and connects Lužec nad Vltavou and Bukol, was constructed using prefabricated UHPFRC, which allowed the development of a thin section, lightweight and durable structure.

Material properties The main material used for the bridge was UHPFRC with steel fibres and was characterised as class C110/130. The material was developed at the Klokner Institute of the Czech Technical University in Prague and its superior mechanical performance and durability were utilised to design the thin deck of the bridge so that the surface of the structure can be walked on without additional waterproofing or covering surfaces. In addition, the footbridge mass was reduced by means of 20 large polystyrene blocks [11].

Construction process The footbridge is a suspended structure with one pylon and two fields with spans of approximately 30 m on one side and 100 m on the other. The width of the deck is 4.5 m and the passing space is 3.0 m. The bridge was designed for a capacity up to 3.5 tonnes (Figure 4.10).

The deck is anchored using 24 pairs of steel cables in the main span and another 7 pairs of cables on the other smaller side.

The bridge deck forms a high-rise arch with a radius of 777 m and is made of precast UHPFRC segments [10]. The UHPFRC panels are reinforced by a pair of ribs located at about one-quarter of the length of the panel from each end [12].

Figure 4.10 The footbridge in Lužec nad Vltavou (Photo credits: Martin Vrut, this file is licensed under the Creative Commons Attribution 3.0 Unported)

Each panel contains an embedded steel weldment at a distance of 1.25 m from the end of the panel, which is used for the anchorage of the hanging rope [12]. The total length of the UHPFRC panels is 5.65 m and the end panels of the deck are accompanied by a monolithic part for connection of the bridge deck and abutments, and for the installation of bridge bearings and the expansion joint on the opposite side of the bridge [12].

Concluding remarks The bridge in this case study demonstrates the efficiency of the use of UHPFRC precast segments for the construction of bridges. The superior performance of the material was utilised for the realisation of the architectural design of the footbridge by using lightweight and durable precast panels.

4.4.2.2 Repair of a bridge deck using UHPC overlay

Introduction and project description This project is focused on the rehabilitation of an existing concrete bridge deck in Socorro, New Mexico, USA [13]. This is a four-span multi-cell box-girder bridge with three intermediate supports consisting of cylindrical columns [13] (Figure 4.11). The bridge has two separate traffic lanes and the surface has a tined finish to provide the required skid resistance.

The existing deck was damaged in various places; these were concentrated at the regions of the negative bending moments where transverse cracks of approximately 1.59 mm width appeared (Figure 4.12).

The main aim of this project was to replace the deteriorated concrete of the deck with a 25 mm thick UHPFRC layer [13].

Material properties The UHPFRC mix design of this study was based on previous research studies at New Mexico State University funded by Transportation Consortium of South-Central States (Tran-SET) [13]. The sand, cement and fly ash were obtained from local sources.

Figure 4.11 The bridge in Socorro, New Mexico, USA [13]

Figure 4.12 Transverse cracking of the bridge decks [13]

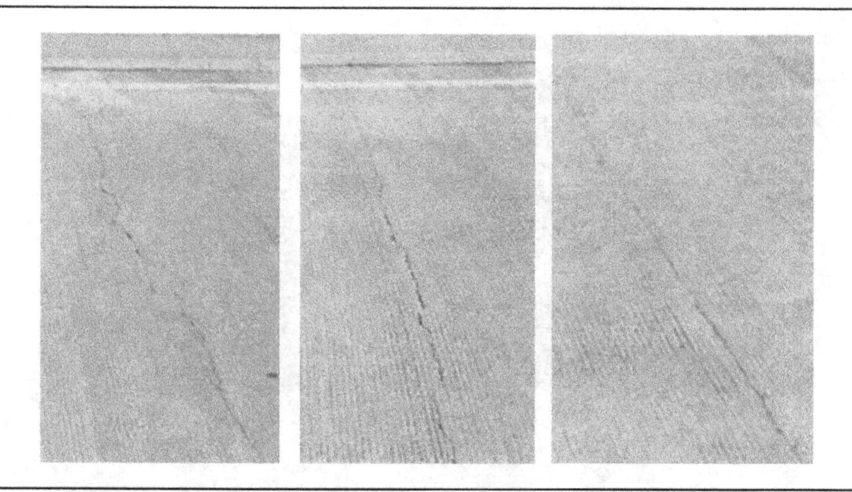

The mix design is presented in Table 4.2. Straight steel fibres of length 13 mm and a length to diameter aspect ratio of 22 were used at a volume fraction of 1.5%.

The average 28-day compressive strength of the examined mix was found to be 134 MPa [13].

Construction process The first step of the process involved the removal of the existing damaged concrete surface (Figure 4.13) together with the expansion joint seals, which were replaced with new ones (Figure 4.14).

For the construction of the new layer, several mock-up castings were scheduled aiming to train the contractor on the mixing and casting processes of UHPFRC. After various mock-up castings, the most suitable mixing and construction process was identified.

Table 4.2 Mixture proportions for UHPC [13]

Materials	Weight: kg/m^3
Cement	737
Silica fume	115
Fly ash	68.8
Sand	1127
Fibres	117
Superplasticiser	42.4
Water	138

Figure 4.13 Removal of deteriorated concrete [13]

(a)

(b)

Figure 4.14 Replacement of the existing expansion joints with new ones [13]

After the removal of the deteriorated deck, the deck was cleaned of debris and contaminants and the texture of the surface was treated. More specifically, the surface was ceramic bead blasted and kept saturated until the casting of the layer [13] (Figure 4.15).

Surface treatment was applied to expose the fine aggregates and then the surface was cleaned and kept saturated until the casting of the overlay. A typical photograph of concrete surface roughening, which is used when additional concrete overlays are applied, is presented in Figure 4.16.

Figure 4.15 Treatment of the surface using ceramic bead blaster [13]

Figure 4.16 Treatment of the surface for the exposure of fine aggregates

For the casting of the UHPFRC overlay, a total of 105 batches were placed, in four placement stages. In the first placement, the UHPFRC overlay was not covered with appropriate plastic sheets after the casting, which led to cracking of the surface (Figure 4.17) caused by drying shrinkage.

In the following three production stages, the curing process was revised.

For the casting of the overlay, the UHPFRC was transferred with buggies and the UHPFRC was placed on the deck manually using shovels (Figure 4.18).

A handled concrete vibrator was used to consolidate the material and a vibrating screed was used to control the thickness of the overlay (Figure 4.19) [13].

Figure 4.17 Cracking of the UHPFRC surface caused by drying shrinkage [13]

Figure 4.18 Casting of the UHPFRC overlays [13]

(a) (b)

Figure 4.19 (a) Hand-held vibrator for UHPFRC consolidation and (b) vibrating screed for the completion of the UHPFRC overlay [13]

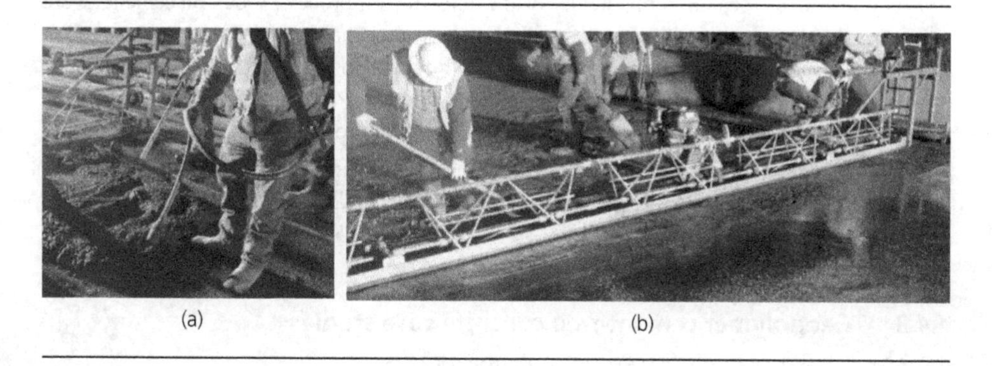

After this stage, a curing compound was applied to the surface (Figure 4.20(a)) and then the surface was covered with a plastic sheet (Figure 4.20(b)) to control the curing and the shrinkage of the UHPFRC layer.

Various in situ measurements were conducted after the end of the construction process including the evaluation of the UHPFRC overlay-to-substrate bond. Non-destructive techniques were applied to identify potential delamination (for example, ground penetrating radar and infrared thermal imaging) while direct tension pull-off tests were also conducted to measure the interfacial adhesion. The average bond strength was found to be 1.65 MPa but this is considered as a conservative value since the premature failure occurred at the resin (which was applied for the testing) to UHPFRC interface [13].

Figure 4.20 (a) Application of the curing compound and (b) application of a plastic sheet to control UHPFRC shrinkage [13]

Concluding remarks This project focused on the replacement of an existing deteriorated bridge deck surface with UHPFRC overlay. An appropriate UHPFRC mix design was used based on previous research in this field. For the construction process, mock placements were conducted to identify the critical issues with the application process, and during these mock castings, it was found that it is important to control the curing of the UHPFRC overlay so as to prevent extensive cracking due to drying shrinkage. The construction process started with the removal of the deteriorated concrete layer followed by the appropriate treatment of the existing surface which was kept under saturated conditions. Next, the concrete was cast and vibrated, and a curing compound was applied. Then the fresh overlay was covered with a plastic sheet. The bond between the UHPFRC and the existing overlay was evaluated by means of non-destructive and pull-off tests.

4.4.3 Geopolymer cement-free concrete case studies

4.4.3.1 Glass-fibre-reinforced geopolymer bridge

Introduction and project description This project describes the development of a new technology for the construction of a bridge using cement-free concrete reinforced with glass-fibre-reinforced polymer (GFRP) bars aimed at the production of a '100 years maintenance-free pedestrian bridge'. This project is the outcome of the successful collaboration of Austeng with Rocla Australia, City of Greater Geelong, Deakin University and Inconmat (Austeng).

Material properties For the geopolymer concrete, commercial mixes from Rocla in Australia were used and an experimental study was conducted to identify the most appropriate material properties and design process [14]. The main constituents were fine and coarse aggregates, fly ash, slag, sodium hydroxide pellets (NaOH), and sodium silicate solution (Na_2SiO_3) [14].

The 28-day compressive strength of the geopolymer mixes were found to be 60 and 62 MPa. With regard to the reinforcement, GFRP bars with ultimate tensile strength equal to 1000 MPa, modulus of elasticity equal to 66.4 ± 2.5 GPa, ultimate strain 1.51% and 25.4 mm diameter were used, while the use of carbon FRP and steel bars were also examined for comparison. Beams were tested under four-point bending and the superior performance of GFRP reinforced geopolymer was highlighted. In addition, some useful conclusions about the design process were derived [14].

Construction process Two of these bridges have been constructed so far for Seagull Paddock in North Geelong. Some initial testing of the GFPR reinforced beams began in May 2019 and these tests show that the geopolymer concrete beams with carbon or glass fibre reinforcing are 15% stronger than traditional steel RC (City of Greater Geelong).

The bridges are presented in Figure 4.21 and they consist of two beams in addition to the precast segments for the deck.

Concluding remarks The bridges examined present a novel construction type whereby the geopolymer cement-free concrete is combined with GFRP bars which have exceptional mechanical characteristics and enhanced durability. This construction type is designed to last

Figure 4.21 Different views of the Geelong GFRP reinforced bridges (Photo credits: Austeng)

(a)

(b)

over 100 years without maintenance, while, at the same time, providing significant environmental benefits through the use of cement-free concrete. This example shows a vision into the future of precast sustainable construction with enhanced durability and structural performance.

4.4.3.2 Geopolymer precast floor panels

Introduction and project description This project involved the use of geopolymer precast panels for the building of the new Global Change Institute (GCI) in Australia (Figure 4.22).

The GCI is an Australian organisation within the University of Queensland that focuses on global sustainability issues including population growth and climate change [15]. The design of the new GCI $32 million (AUD) building sets various parameters and is specifically focused on key sustainability factors such as having zero net carbon emission for building operation, and being carbon neutral with carbon offset [15]. Various options and materials were considered before the project consultants selected the use of geopolymer concrete as the preferred option for the precast floor beams used in the three suspended floors [15].

Figure 4.22 The new GCI building (Photo credits: Angus Martin)

Material properties All tests were undertaken by Wagners and the results were assessed by Dr J. Aldred from Aecom [15, 16]. The binder of the examined geopolymer concrete consisted entirely of supplementary cementing materials. Although there was the expectation that there would be a slow strength development at 7 days, the strength was above 80% of the 28-day strength. Measurements were taken over a six-month period and all the results show a 28-day strength above the 40 MPa requirement [16]. The respective indirect tensile strength value was found to be equal to 4.8 MPa, which is equivalent to a direct tensile strength of 4.3 MPa [16]. Detailed testing of the modulus of elasticity, Poisson's ratio, shrinkage, creep and durability of the mix were also performed and the detailed results are available in a relevant study [16].

Construction process Geopolymer concrete was used for the 33 precast floor beams (320 m^3) that formed three suspended floors in the building [15]. The specification for the geopolymer concrete in these beams relied on specific requirements and detailed testing [16]. In addition to the design and specification details, the geopolymer concrete was required to meet all of the handling requirements involved in batching and delivering concrete and these were demonstrated by Wagners [15].

The first precast floor beam panel constructed was used as the prototype, which was tested (Figure 4.23). The results showed that the measured deflection under a uniformly distributed load of 10 tonnes was equal to 2.85 mm, which was less than the predicted value of 3 mm [15].

The construction of the four-storey building was completed with the installation of the 33 panels that were used for the construction of the three suspended geopolymer concrete floors. The installation of one of the 33 panels is presented in Figure 4.24.

Concluding remarks This study shows the potential of geopolymer concrete for multi-storey structures. The use of geopolymer precast segments was found to be the optimum solution for this structure after considering other construction types and materials. This technique should

Figure 4.23 Full-scale testing of the prototype (Photo credits: Wagners Concrete and Precast Concrete) [15]

Figure 4.24 Installation of one of the precast panels (Photo credits: Wagners) [17]

be further exploited for the development of a technology that could be used extensively in residential buildings and in load-bearing precast geopolymer concrete elements.

4.5. Strengthening using novel high-performance fibre-reinforced cementitious materials

In this section, a collection of strengthening applications using high-performance and geopolymer concrete are described. The main findings of experimental investigations on the performance of RC beams strengthened with additional layers of UHPFRC will be presented first and then the results of additional studies about the UHPFRC-to-concrete interfaces and the 'size effect' of the thickness of UHPFRC layers (Section 4.5.1) are given. The performance of strengthened beams using fibre-reinforced geopolymer concrete has also been studied and is presented in Section 4.5.2. Finally, the potential use of UHPFRC for the strengthening of unreinforced masonry (URM) is examined and some preliminary results are presented in Section 4.5.3.

4.5.1 Strengthening of existing RC beams using UHPFRC [18]

Strengthening of RC structural elements is an emerging field as this can allow the reuse of existing structures. The majority of the existing RC structures are either designed with old or without code provision, or they are damaged and, therefore, they need to be repaired and strengthened.

The selection of the appropriate materials for the strengthening applications is a complex process and various parameters need to considered depending on the application and on the exact requirements.

Additional reinforced concrete layers and jackets were traditionally used for the strengthening of RC beams and columns. In the last few years, the application of novel high-performance materials for strengthening applications has been extensively studied.

In this section, the results of an experimental study on RC beams strengthened with additional UHPFRC layers with and without additional steel rebar reinforcement are presented. Comparisons are also made with respective beams strengthened with conventional RC layers. Large-scale RC beams were examined and strengthening layers with various characteristics were applied. The beams were tested with the application of a load to the middle of the span, and the load against mid-span deflection was monitored in addition to the slip at the old-to-new concrete interfaces. The results were critically evaluated and valuable conclusions for the effectiveness of the strengthening techniques are presented.

4.5.1.1 Description of the examined specimens and experimental results of strengthened RC beams

The experimental results are based on two previous experimental works on the structural strengthening of RC beams with conventional RC [19] and UHPFRC [20] layers.

The specimens investigated in these two studies [19, 20] had relatively similar geometry and loading set-ups, which allows a direct comparison and a critical evaluation of the two methods.

Beams strengthened with conventional RC layers [18] In the first of these two studies [19], an experimental investigation on the flexural strengthening of RC beams using additional RC layers was conducted.

The original RC beam (Figure 4.25) had a length of 2200 mm and a 150 mm by 250 mm rectangular cross-section. The beams were strengthened with 212 B500 steel with a cover of 25 mm on their tensile side (Figure 4.25(a)). As shown in Figure 4.25, stirrups with a diameter of 8 mm and spacings of 100 mm and 50 mm were positioned to provide shear reinforcement.

The concrete of the initial beam was found to have a compressive strength of 39.5 MPa. A new concrete layer of 50 mm thickness, reinforced with 212 B500 steel, was added for the purpose of strengthening. Two strengthened beams were investigated: one with an interface roughened

Figure 4.25 (a) Initial and (b) strengthened specimens using RC [18]

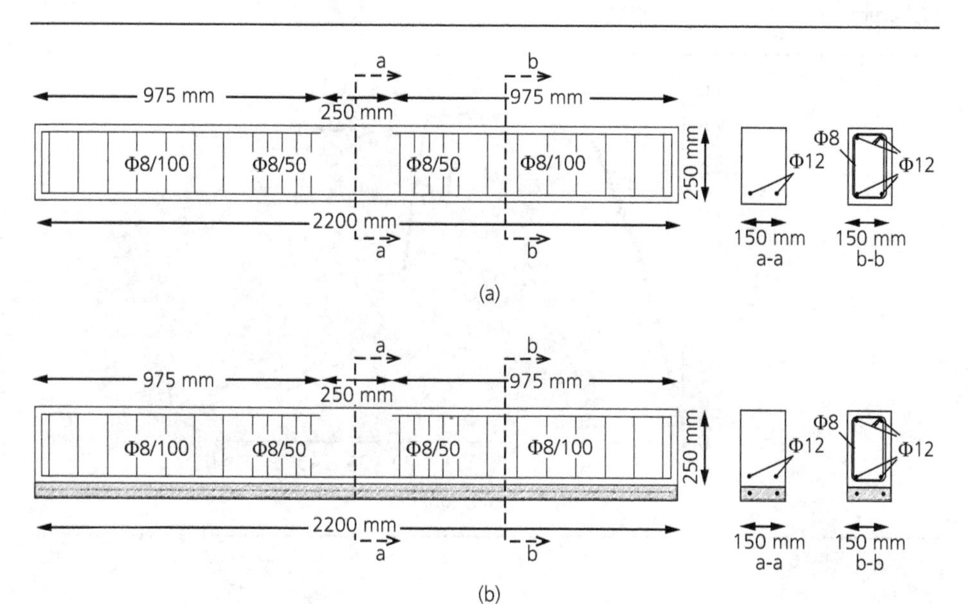

to a depth of 2 to 3 mm (T1) and the other with an interface roughened to a depth of around 1 to 2 mm (T2). For specimens T1 and T2, the compressive strength of the concrete of the new layers was found to be 38.9 MPa and 34 MPa, respectively.

Figure 4.26 depicts the loading conditions. The longitudinal dimension of the structural member was 2000 mm, while the distance between the two loading points (s), was 500 mm.

Figure 4.27 illustrates the load against deflection results of the beams.

Figure 4.26 Loading conditions of the examined beams [18]

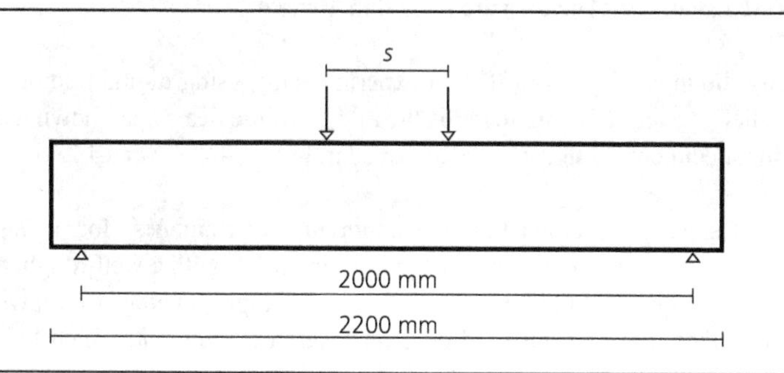

Figure 4.27 Graph of load–deflection for the strengthened beams with the use of conventional RC layers [19]

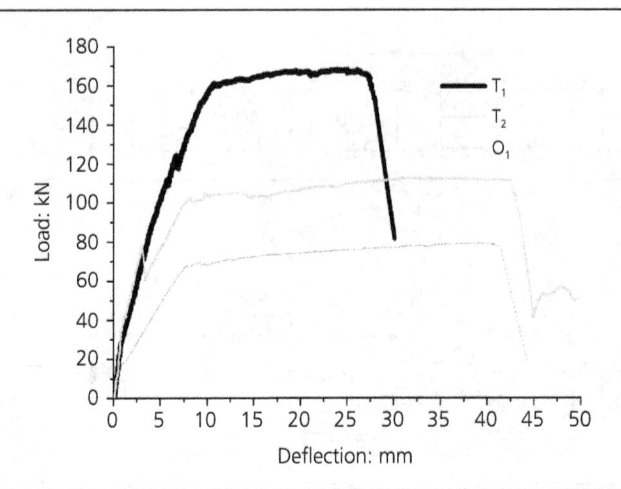

The results of Figure 4.27 demonstrate a significant increase in the stiffness of RC beams, as expected, with a substantial enhancement of the ultimate load capacity for both specimens, T1 and T2. However, specimen T2 yielded to premature collapse due to a smaller degree of roughness at the interface.

The mode of failure of the studied specimens is illustrated in Figure 4.28 at a loading stage that is close to the maximum load capacity of Figure 4.27. It is clear from the failure patterns of the three examples under investigation (Figure 4.28) that all of the specimens exhibited flexural cracks in varying degrees, with some shear cracks developing later after the main flexural cracks had already formed. Additionally, specimen T2 had delamination of the new concrete layer, which is seen in Figure 4.28(c) and relates to a reduction in the load capacity of this specimen, as shown in Figure 4.27.

Using the findings of Figure 4.27 as a guide, the load and deflection values for the characteristic sites of yield (P_y) and failure (P_u) as well as the initial stiffness (K) at 3 mm mid-span deflection were calculated. The data are shown in Table 4.3.

The slip along the interface was evaluated experimentally using digital micrometres at each support, and then at every 330 mm towards the middle of the beams, as shown in Figure 4.28. The results for specimens T1 and T2 are displayed in Figure 4.29(a) and 4.29(b), respectively.

The results of Figure 4.29 demonstrate the importance of roughness for the structural performance of the examined beam. In the case of specimen T1, with a well-roughened interface the slip values were quite small and they reached up to approximately 1 mm, whereas in the case of T2, interface failure occurred due to poor surface treatment, which led to high slip values of approximately 8 mm and subsequent delamination.

Figure 4.28 Illustration of the failure mode of specimens (a) O1, (b) T1 and (c) T2 for the stage of the ultimate load capacity [18]

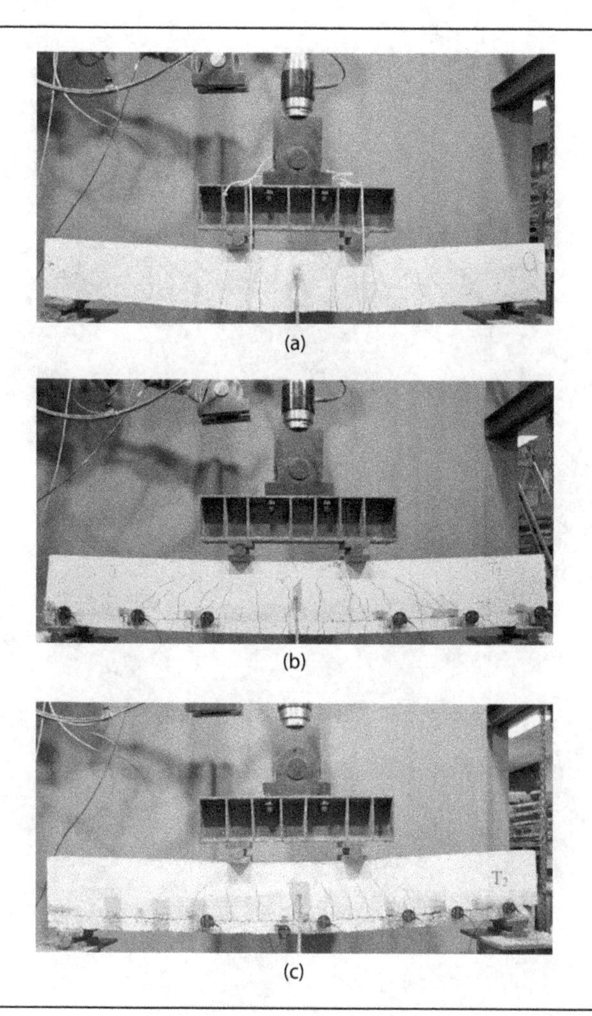

Table 4.3 Results for selected characteristic points [18]

	T1	T2	O1
P_y: kN	160	101	68
Δy: mm	10.5	8	8
P_{max}: kN	167	113	79
δP_{max}: mm	26.5	42	39
K: kN/mm	23	26	11

Figure 4.29 Slip distribution along the interface for the maximum load capacity for (a) T1 and (b) T2 [18]

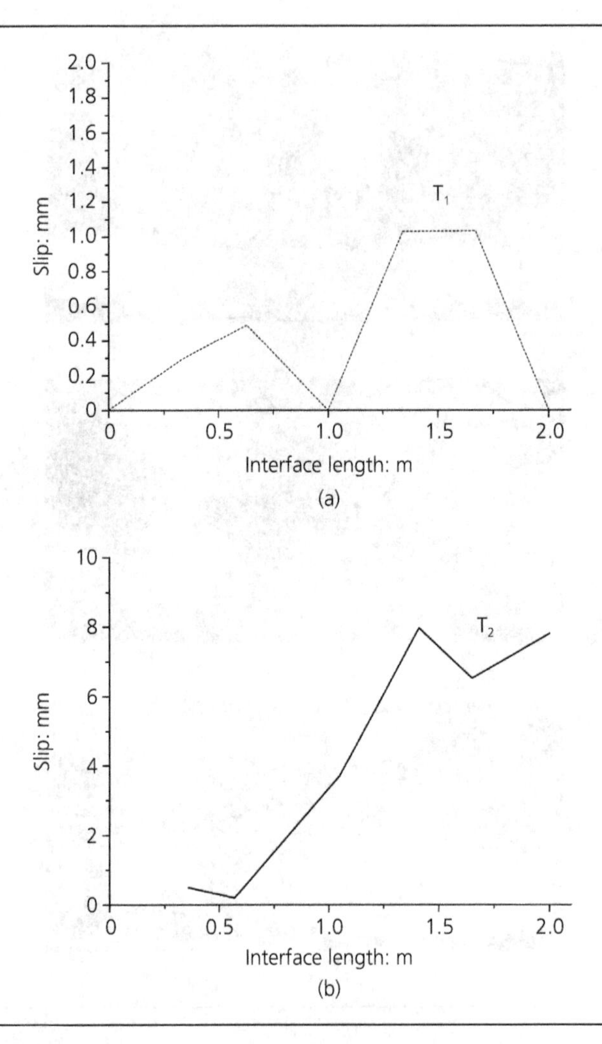

Beams strengthened with UHPFRC layers [18] The results for elements reinforced with UHPFRC layers are described in this section. Two beams possessing similar geometry and properties to those addressed in the previous section were investigated and strengthened with UHPFRC layers [20] (Figure 4.30). Both initial beams (P) had two longitudinal ribbed steel bars of 12 mm diameter and 2150 mm length attached to the tensile side. Stirrups of diameter 10 mm and with 150 mm spacing were added as shear reinforcement.

The examined specimens were strengthened by the addition of UHPFRC layers with 50 mm thickness along their entire length (Figure 4.30). To this end, the initial beams were roughened to a depth of 2–2.5 mm, which was a similar degree of roughness to that used in specimen T1. Accordingly, two specimens incorporated no steel bars in the layer (U), whereas two other

Figure 4.30 (a) Initial and (b) strengthened beams with the addition of UHPFRC layers [18]

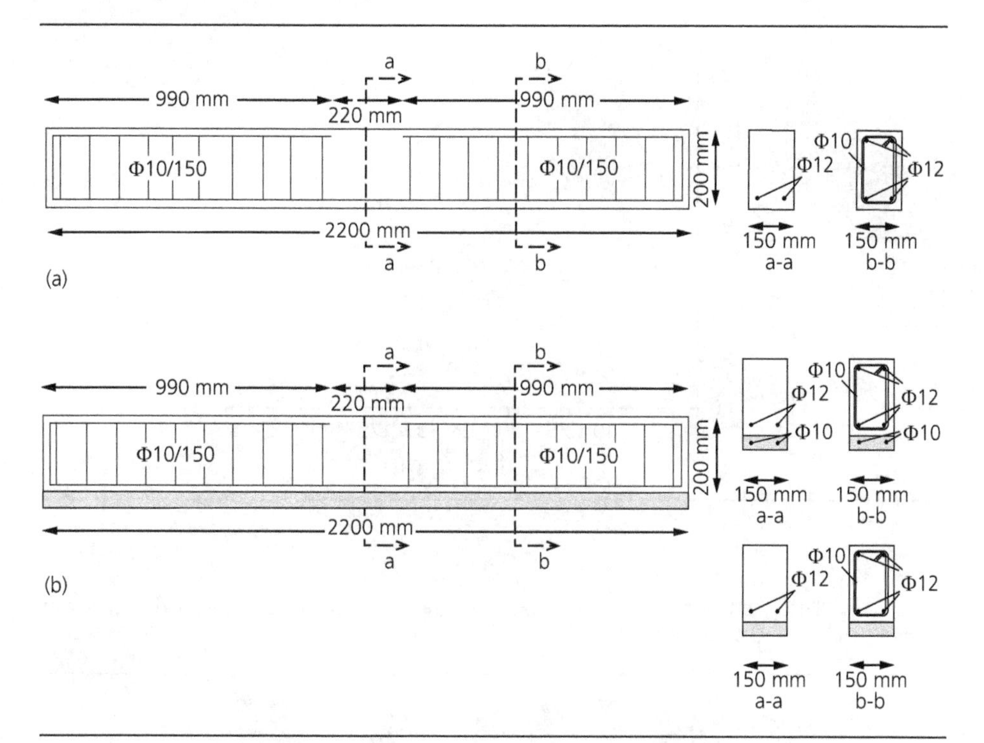

specimens included two ribbed steel bars measuring 10 mm in diameter made from B500 steel (UB). Plastic spacers were implemented for both the initial beam and the UHPFRC layers to guarantee a concrete cover of 25 mm. The compressive strength of the concrete in the initial beam was identified as 30.9 MPa, whereas the UHPFRC demonstrated a compressive strength of 136.5 MPa and a tensile strength of 11.5 MPa at the time of testing [20].

The loading set-up is illustrated in Figure 4.26, with 400 mm spacing between loading points. Two specimens were evaluated for each type and the average curves are illustrated in Figure 4.31.

Figure 4.31 demonstrates that the addition of a UHPFRC layer without steel bars (U) leads to significant enhancement of the initial stiffness. In terms of maximum load capacity and post-peak behaviour, the response of the specimens strengthened with a UHPFRC layer without steel bars (U) was comparable with the original beams (P). However, UHPFRC strengthened with steel bars (UB) resulted in significantly enhanced load capacity as well as improved stiffness due to the presence of longitudinal steel bars. Figure 4.32 presents failure modes for specimens near the maximum loading stage.

The results of Figure 4.32 show typical flexural failure modes in all the examined specimens. It is evident from Figure 4.32(b) that a major crack formed at the tensile side of U specimens

Figure 4.31 Graph of load–deflection for beams strengthened with UHPFRC layers [20]

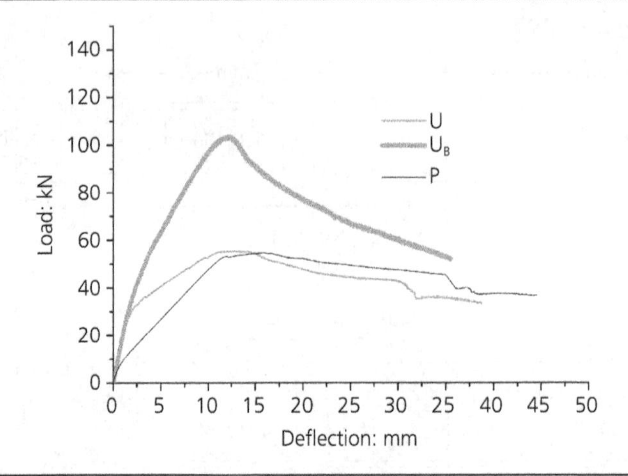

Figure 4.32 Illustration of typical failure modes for the stage of the ultimate load capacity of specimens (a) P, (b) U and (c) UB [18]

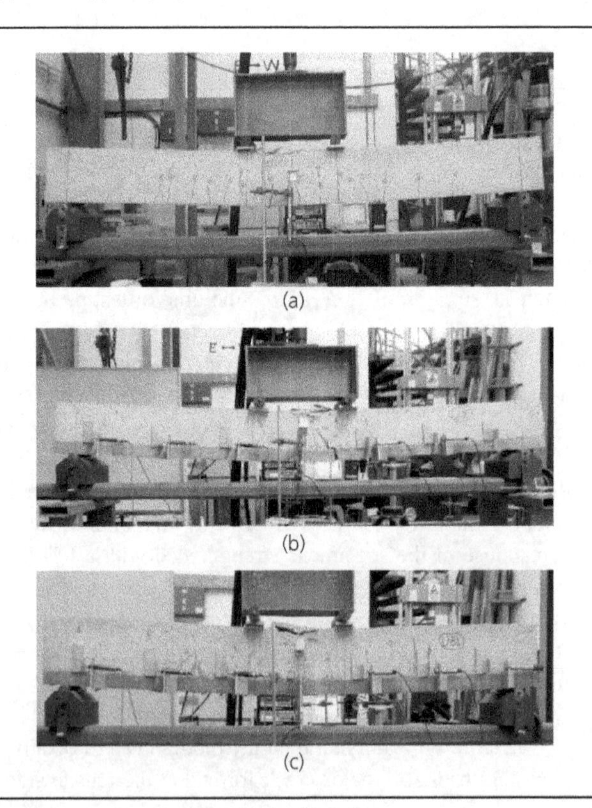

where no steel bars were added, which had a detrimental effect on the ultimate load capacity. The results show an evident change in the initial slope of the load–deflection curve due to the initiation of cracking. The yield points (P_y, δy) and failure points (P_u, δu) as well as stiffness (K) at 3 mm mid-span deflection were determined and the results for the ultimate load (P_{max}), the respective deflection (δ_{Pmax}) and the stiffness (K) are presented in Table 4.4.

Slip along the interface was recorded experimentally using linear variable differential transformers (LVDTs), beginning at a distance of 100 mm from the supports, and then again at 300 mm intervals up to the midpoint of the beams, as depicted in Figure 4.32. The results are shown in Figure 4.33 for specimens U and UB.

The results of the slip distribution along the interface (Figure 4.33) show maximum slip values of less than 0.4 mm in both U and UB specimens, which indicated very good connection conditions between the UHPFRC and the existing concrete substrate.

Comparisons of the effectiveness of the use of UHPFRC and RC layers for the strengthening of existing RC beams [18] In this area, the effectiveness of the two examined strengthening techniques, with conventional RC and with UHPFRC layers, is evaluated. Results for stiffness and ultimate load capacity are compared and the interface conditions are analysed based on the slip findings. Figure 4.34 displays the increase in the amount of stiffness in relation to their original beams.

The results from Figure 4.34 show an increase in stiffness of between 100% and 150% when specimens with UHPFRC layers reinforced with steel bars (specimen UB) are used. The peak stiffness increment was recorded at 150%, for specimen UB.

The specimen with UHPFRC only (U) had 100% increase in stiffness, whereas the increment for the specimens strengthened with RC layers was found between 109% and 136%.

The results of the maximum load are illustrated in Figure 4.35. These results show a negligible increment in the ultimate load capacity when UHPFRC without steel bars is used (U). This is attributed to the formation and opening of cracks at the UHPFRC layer, with subsequent flexural strength reduction (Figure 4.31). The load–deflection results (Figure 4.31) show that the ultimate load of the U specimens approaches the respective load capacity of the initial beam (P). Also, it can be concluded that the addition of UHPFRC layers result in a significant ultimate load enhancement of approximately 90% compared with the respective results for the specimens with UHPFRC layers without steel bars (U). The specimens strengthened with

Table 4.4 Selected experimental results for the examined specimens [18]

	U	UB	P
P_{max}: kN	55.34	103.49	54.55
δ_{Pmax}: mm	12.26	12.23	15.88
K: kN/mm	12	15	6

Figure 4.33 Distribution of slip at the interface of (a) U and (b) UB for the ultimate load capacity [18]

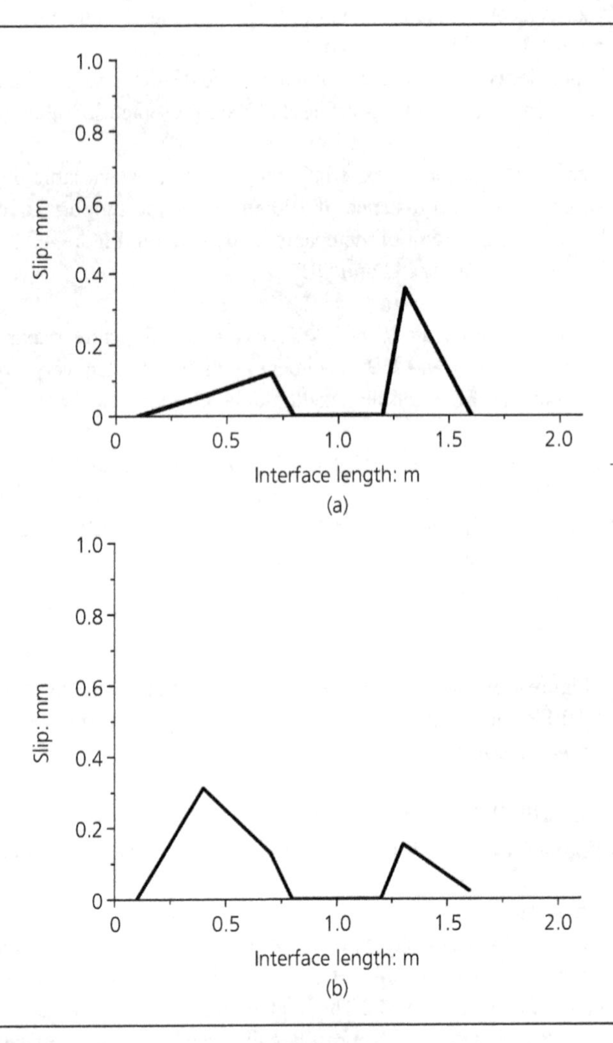

conventional RC layers show a 43% and a 111% ultimate load enhancement for interfaces with insufficient treatment (T2) and for well-roughened interfaces (T1), respectively. The steel bars had 12 mm diameter in cases T1 and T2, and 10 mm for UB.

Therefore, to be able to perform some direct comparisons, the maximum load of specimens T1 and T2 were adjusted by multiplying the increment of the maximum load capacity with the ratio of the cross-sectional area of 10 mm to 12 mm steel bars (T1 adj, T2 adj). In this way, the maximum load capacity of the examined specimens could be compared. The results show that a 90% load increment was achieved in the case of specimen UB (Figure 4.35), whereas the respective results of specimens T1 adj and T2 adj were found to be equal to 77% and 30%, respectively.

Figure 4.34 Comparisons of the stiffness enhancement [18]

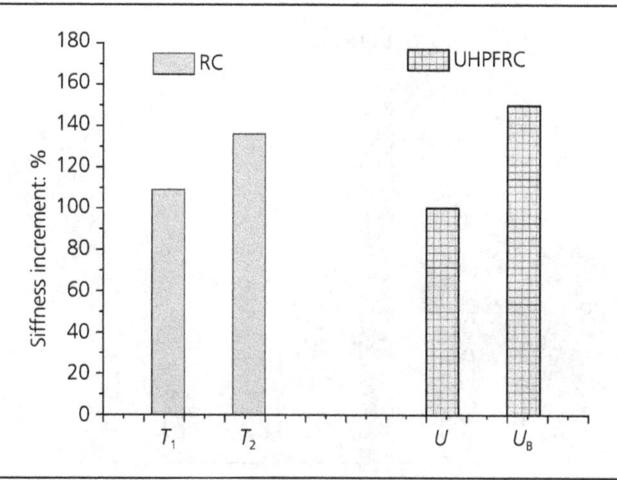

Figure 4.35 Ultimate load enhancement results [18]

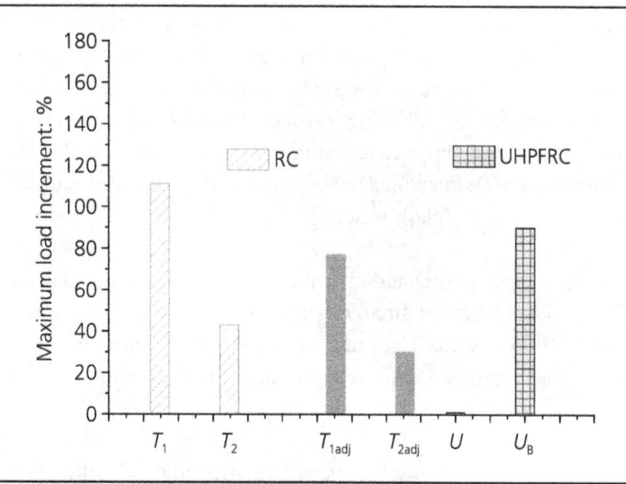

Figure 4.36 shows the maximum interface slip along the interface for the stage of the maximum load capacity for all the examined specimens.

Figure 4.36 shows that the interface slip values in both specimens U and UB are much lower than those in the T1 and T2 specimens, which demonstrates the improved interface connection of the UHPFRC layers in comparison with the conventional RC layers. More specifically, for specimens strengthened with RC layers with well-roughened (T1) and poor interface conditions (T2), the maximum slip was 1 mm and 8 mm, respectively, whereas the values for

Figure 4.36 Comparisons of the maximum interface slip results [18]

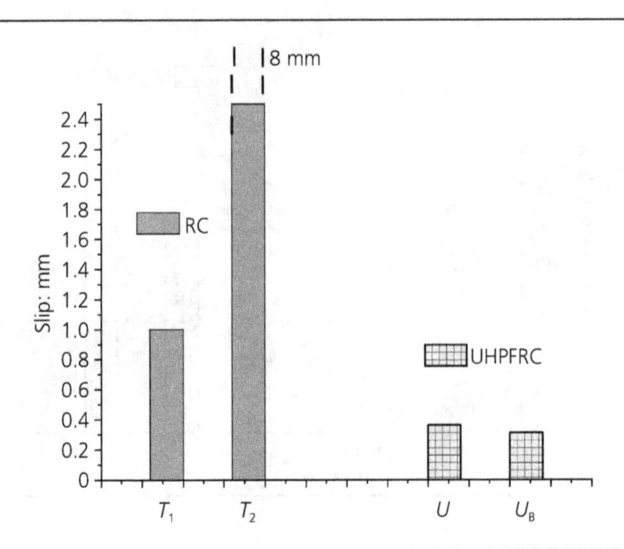

strengthened beams with UHPFRC layers were found to be 0.36 mm and 0.31 mm for specimens U and UB, respectively. The improved interface conditions for specimens reinforced with UHPFRC layers could also be seen visually in the failure modes (Figure 4.32), where there were no interface cracks visible during loading. When RC layers were used to strengthen the existing beams, cracks and slippage at the interface could be visually observed from the experimental failure modes (Figure 4.28). The enhanced performance of the UHPFRC-to-concrete interfaces is studied in detail in Section 4.5.1.2.

Concluding remarks This experimental investigation focused on the evaluation of the effect of RC and UHPFRC layers for the strengthening of existing RC beams. Layers with conventional RC and UHPFRC were used, and the effect of the addition of steel bars in the UHPFRC layers was also examined. The main findings of the experimental investigation are as follows.

- The lowest slip values were found for specimens strengthened with UHPFRC layers, which were below 0.4 mm in both of the examined specimens, whereas the values for specimens strengthened with RC layers were found to be much higher, at around 1 mm and 8 mm for well-roughened and poor interface conditions, respectively.
- The specimens with the highest stiffness and ultimate load capacity enhancement and the lowest interface slip values were those strengthened with UHPFRC layers and steel bars.
- The addition of UHPFRC layers led to significant enhancement of the initial stiffness of strengthened beams. In the case of the specimen with the UHPFRC layer without steel bars, a major crack was developed at the UHPFRC layer, resulting in a significant loss of the strengthened element's initial stiffness and a subsequent reduction of the ultimate

load and post-peak capacity. The ultimate load capacity of this specimen was not significantly higher that for the beam without strengthening.
- The addition of RC layers led to significant enhancement of stiffness and maximum load capacity of the beam with the well-roughened interface. However, the strengthened specimen with insufficient interface treatment showed premature failure and reduced ultimate load capacity compared with the strengthened beam with the well-roughened interface.

4.5.1.2 Experimental investigation of the concrete-to-UHPFRC interfaces [21]

One of the most widely employed procedures globally, particularly in earthquake-prone areas, is the strengthening of existing RC elements with additional concrete layers or jackets. The improvement of the stiffness, strength and ductility of load-bearing structural RC elements frequently involves the use of traditional RC layers and jackets. Due to the enormous advantages associated with the improved mechanical qualities and durability characteristics, the use of novel high-performance materials, in particular the use of additional layers of UHPFRC, is an area that is rapidly expanding. The superior tensile and compressive strength of UHPFRC and the significantly enhanced post-cracking energy absorption are the main reasons for the selection of UHPFRC as a strengthening material. These outstanding mechanical properties, in addition to its enhanced durability, can offer significant benefits for the structural upgrade of existing RC beams and/or columns.

The efficiency of the approach has been the subject of much research [20, 22–27]. Precast UHPFRC strips were used by Farhat *et al.* [22] and by Al-Osta *et al.* [25], and comparisons with cast in situ UHPFRC have also been published [25]. However, cast in situ UHPFRC is the subject of the majority of the studies under consideration, and it was demonstrated that this technology can significantly improve the stiffness and strength of load-bearing structural elements [20, 23, 25–27].

The connection between the UHPFRC and the existing structure is a crucial feature that affects the overall performance of the strengthened elements because failure at the interface and debonding of the new layer or jacket are the most common reasons for failure in cases of elements strengthened with this technique. The preparation of the surface and assessment of the bond strength between UHPFRC and concrete are two of the most crucial factors for the successful application of this process. UHPFRC-to-concrete interfaces appear to have increased strength [20]. It has also been proved that the addition of mechanical connectors (for example, dowels) has a positive effect that leads to enhanced connection and improved resistance to cracking for the beams strengthened with UHPFRC layers [27]. Although there are published studies on the application of UHPFRC layers and jackets, there is very limited research on the bond strength between UHPFRC and conventional concrete.

In this section, the results of an extensive experimental study are presented [21] in addition to a critical evaluation and discussion about the performance of the UHPFRC-to-concrete interfaces. The shear stress–slip behaviour of different types of UHPFRC-to-concrete interfaces was examined and comparisons with conventional concrete-to-concrete interfaces are also presented [21].

Description of the specimens used for the push-off tests and experimental results In this study, conventional concrete cubes bonded to UHPC (without fibres) and UHPFRC (with 3% steel fibre volume fraction) cubes were examined [28]. Three 100 mm side cube triplet specimens were investigated for the evaluation of the interface conditions (Figure 4.37). The conventional concrete (NC) cubes were cast first and they were cured for 14 days. The surfaces of the conventional concrete cubes were then roughened with a pistol grip needle and they were placed into the appropriate 300 mm by 100 mm by 100 mm moulds. Two conventional concrete cubes were placed into the moulds at a spacing of 100 mm, and then UHPC/UHPFRC was cast between the two conventional concrete samples to create the triplet specimens (Figure 4.37).

UHPC with no fibres and UHPFRC with 3% steel fibres were examined to assess the effect of the steel fibres [28]. Typical concrete-to-concrete interactions were studied. Table 4.5 shows the mix composition for conventional concrete, whereas Table 4.6 shows the mix compositions for UHPC and UHPFRC.

A value of 3% (235.5 kg/m^3) was chosen as a representative amount of a high steel fibre volume fraction for the UHPFRC (Table 4.6).

A needle pistol was used for the roughening of the contact surfaces of the exterior conventional concrete cubes and two different levels of roughness were achieved.. The sand patch method was applied to calculate the degree of roughness. Silica sand was used to cover the roughened

Figure 4.37 Geometry and loading conditions of the triplet specimens [21]

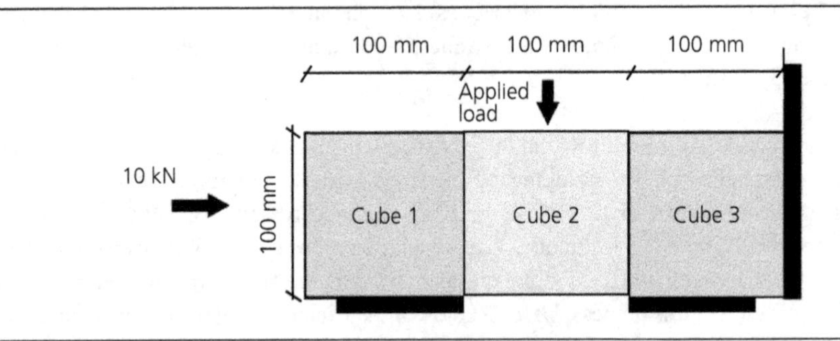

Table 4.5 Conventional concrete mix [21]

Material	Quantity: kg/m^3
Cement	340
Sand	1071
Aggregates	714
Water	205

Table 4.6 UHPFRC and UHPC mix [21]

Material	Quantity: kg/m^3
Cement	620
GGBS	434
Silica fume	1410
Silica sand	1051
Superplasticiser	59
Water	185
Cement	620
Steel fibres	UHPC: 0 UHPFRC: 235.5

surface, and then it was removed and its mass was calculated. The average mass of the first type of roughening (RI) was determined to be 14.67 g, whereas the average mass of the second type (RII) was much higher and equal to 26.92 g. The equivalent thickness, which is shown in Table 4.7, was calculated using the bulk density of silica sand, which is equal to 1730 kg/m^3. The middle cube was cast after the surfaces had been roughened, and the specimens were cured for at least 28 days before testing.

Table 4.8 provides an overview of the contact conditions and the concrete type used in the examined triplet specimens. Three samples were evaluated for each of the examined categories.

Figure 4.38 displays the experimental set-up for the push-off tests [28].

A load of 10 kN was applied normally to the interfaces during testing. The central cube was loaded in the perpendicular/vertical direction using displacement control at a rate of 10 μm/s. The 10 kN normal load varied during the test and it was manually adjusted during the testing to keep it as close as possible to 10 kN [28]. For the monitoring of the interface slip at the two sides of the interface as well as any movement of the specimens at the end, three LVDTs were used. The central cube was loaded until the failure of the triplet specimens. The compressive

Table 4.7 Average roughness thicknesses for the examined specimens [21]

Surface roughness	Abbreviation	Thickness: mm
Smooth	S	0
Rough I	RI	0.85
Rough II	RII	1.56

Table 4.8 Overview of the examined specimens [21]

	Triplet ID	Description of the samples
Concrete-to-concrete	C-C_S	Cube 1,3: NCe Cube 2: NCi Interface: Smooth (S)
	C-C_RI	Cube 1,3: NCe Cube 2: NCi Interface: Rough I (RI)
	C-C_RII	Cube 1,3: NCe Cube 2: NCi Interface: Rough II (RII)
UHPC 0%-to-concrete	C-UHP 0%_S	Cube 1,3: NCe Cube 2: UHPC 0% Interface: Smooth (S)
	C-UHP 0%_RI	Cube 1,3: NCe Cube 2: UHPC 0% Interface: Rough I (RI)
	C-UHP 0%_RII	Cube 1,3: NCe Cube 2: UHPC 0% Interface: Rough II (RII)
UHPFRC 3%-to-concrete	C-UHP 3%_S	Cube 1,3: NCe Cube 2: UHPFRC 3% Interface: Smooth (S)
	C-UHP 3%_RI	Cube 1,3: NCe Cube 2: UHPFRC 3% Interface: Rough I (RI)
	C-UHP 3%_RII	Cube 1,3: NCe Cube 2: UHPFRC 3% Interface: Rough II (RII)

strength of the cubes of all the examined concrete types was evaluated using 100 mm cubes, and the results are shown in Table 4.9.

Figures 4.39, 4.40 and 4.41 [28] show the results of the push-off shear stress against slip together with the average curves. The slip values were calculated using the average of the slip values at the two interface sides, or from only one sensor in the event of undesirable faults of the LVDTs during the testing or local failures on one of the two sides. Additionally, some of the analysed samples experienced undesirable failure types; in these instances, only two of the three specimens were considered in the computation of the average curves [21].

Figure 4.38 Test set-up of the push-off testing of the triplet specimens [21]

Table 4.9 Mean compressive strength of the different concrete types [21]

Concrete type	Compressive strength: MPa
NCe	29
NCi	18
UHPC 0%	99
UHPFRC 3%	105

Figure 4.42 displays a comparison of all of the results for average shear stress against slip.

The findings demonstrate that shear stress values are significantly lower for all of the examined specimens with smooth interfaces (C-C_S, C-UHP 0%_S and C-UHP 3%_S) compared with the corresponding specimens with roughened surfaces. Shear stress values for the conventional concrete-to-concrete samples (C-C_RI and C-C_RII) with rough and very rough interfaces are much lower than for the corresponding UHPC 0%-to-concrete and UHPFRC 3%-to-concrete samples. The UHPFRC 3%-to-concrete interfaces with rough and very rough interfaces (C-UHP 3%_RI and C-UHP 3%_RII) have the highest interface shear stress values.

Additionally, in the majority of the cases under study, the maximum shear stress is activated for very low values of relative slip along the interface (between 0.01 mm and 0.02 mm), which is in agreement with the Greek retrofitting code [29].

The experimental results were compared with the available analytical models for smooth and rough interfaces, and the results are presented in this section [29].

Figure 4.39 Shear stress–slip curves for (a) C-C_S, (b) C-C_RI and (c) C-C_RII specimens [21]

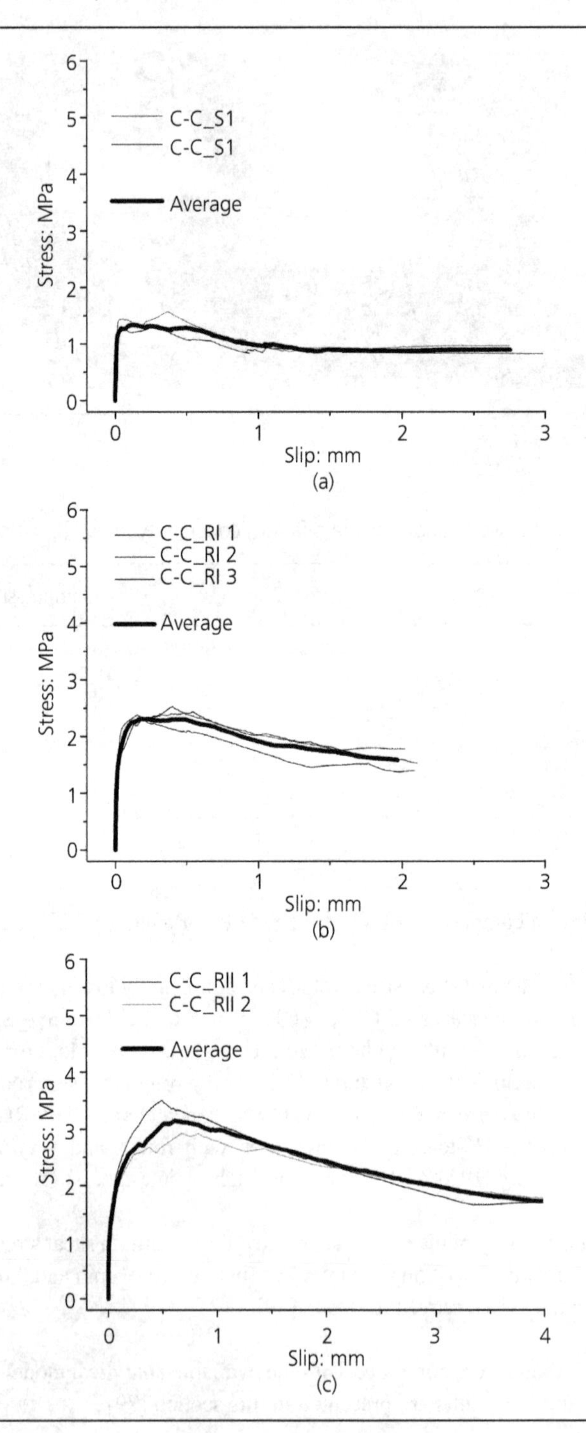

Figure 4.40 Shear stress–slip curves for (a) C-UHP 0%_S, (b) C-UHP 0%_RI and (c) C-UHP 0%_RII specimens [21]

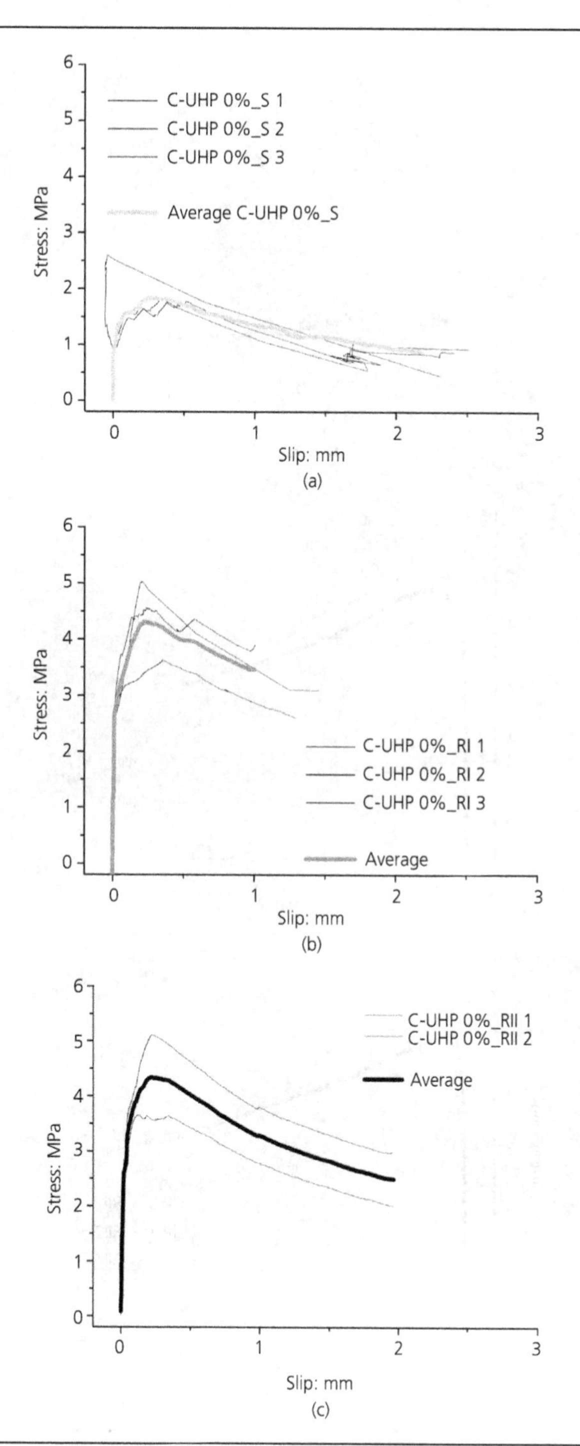

Figure 4.41 Shear stress–slip curves for (a) C-UHP 3%_S, (b) C-UHP 3%_RI and (c) C-UHP 3%_RII specimens [21]

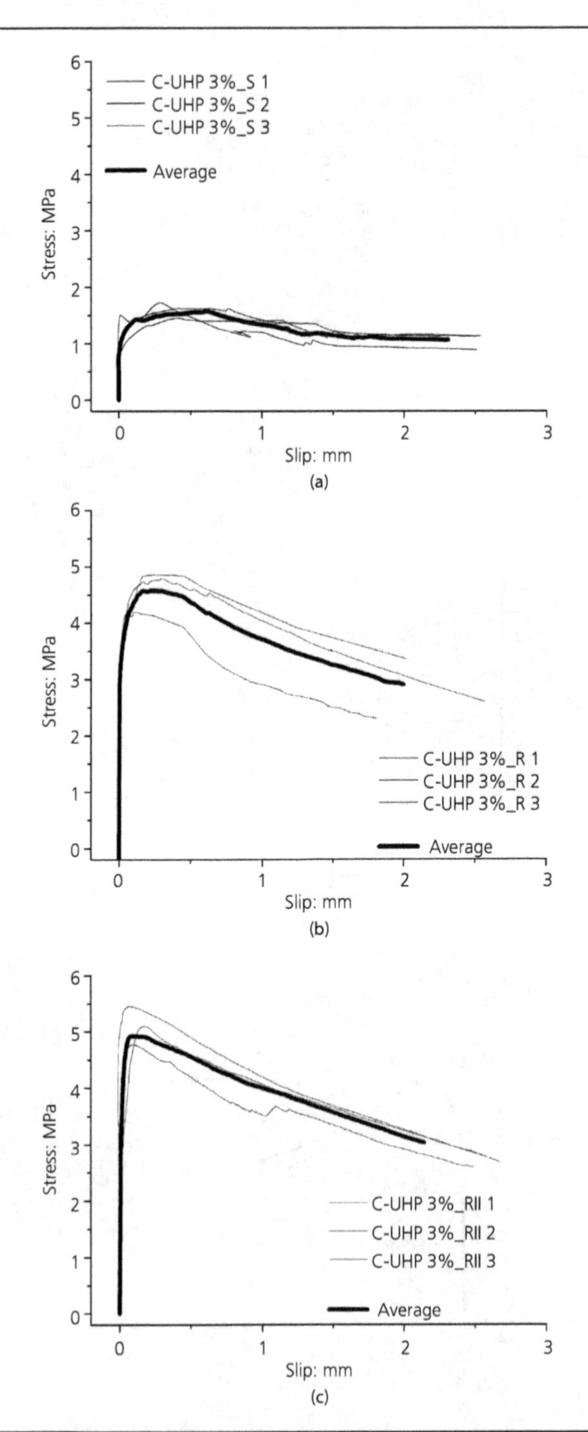

Figure 4.42 Average shear stress–slip curves for all of the examined specimens [21]

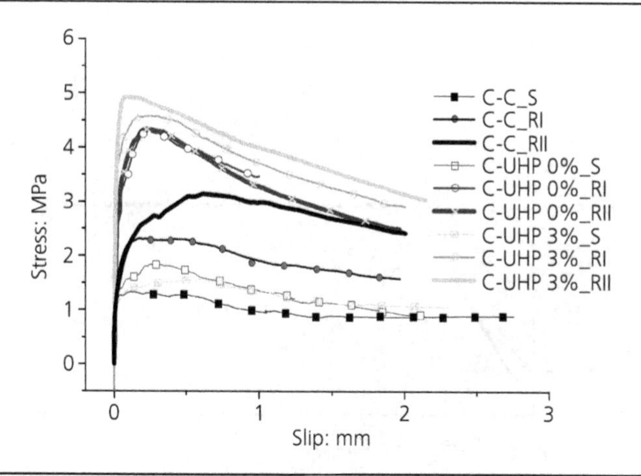

The shear stress against slip analytical models for the design of smooth and rough interfaces are available in the literature, and they are illustrated in Figure 4.43 [29].

Equation 4.2 can be used to obtain the maximum frictional resistance (τ_{fud}) for smooth interfaces (Figure 4.43(a)).

$$\tau_{fud} = 0.4 \times \sigma_{cd} \qquad (4.2)$$

where:

σ_{cd} is the design value for the interface's minimum total normal compressive stress.

The proposed model of Figure 4.43(b) is for rough interfaces. Equations 4.3–4.5 can be used to compute the shear strength (τ) for a range of slip values (s), starting at zero and going up to the maximum slip (s_{fu}), which occurs for the highest frictional resistance (τ_{fud}).

$$\text{For } \frac{s_f}{s_{fu}} \leq 0.5,\ \frac{\tau}{\tau_{fud}} = 1.14\sqrt[3]{\frac{s_f}{s_{fu}}} \qquad (4.3)$$

$$\text{For } \frac{s_f}{s_{fu}} > 0.5,\ \frac{\tau}{\tau_{fud}} = 0.81 + 0.19\frac{s_f}{s_{fu}} \qquad (4.4)$$

Equation 4.5 can be used to calculate the maximum shear stress value.

$$\tau_{fud} = 0.4 \times \left(f_{cd}^2 \times \sigma_{cd}\right)^{1/3} [\text{MPa}] \qquad (4.5)$$

where:

f_{cd} is the design compressive strength of the weaker of the concretes of the interface.

Figure 4.43 Analytical models for the design of smooth (a) and rough (b) interfaces [29]

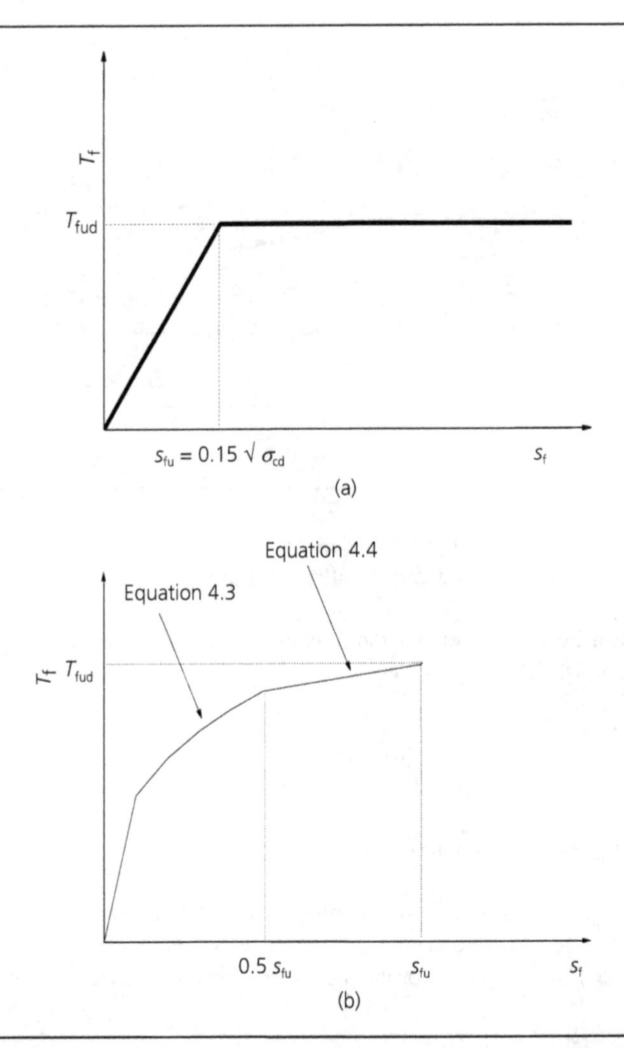

Equation 4.2 was applied to the current study's smooth interfaces with the assumption that σ_{cd} was equal to 1 MPa, which was calculated based on the 10 kN load applied to a 100 mm by 100 mm contact area. Using the compressive strength characteristics of the examined specimens, two different curves were derived for the rough interfaces. More specifically, the lowest compressive strength at the interface for C-C_RI and C-C_RII was experimentally determined to be equal to 18 MPa, whereas the respective compressive strength value for C-UHP 0%_RI, C-UHP 0%_RII, C-UHP 3%_RI and C-UHP 3%_RII was experimentally determined to be equal to 29 MPa (Table 4.9). In Figure 4.44(a) and Figure 4.44(b), the predicted graphs for smooth and rough interfaces are presented alongside the corresponding experimental data.

The findings demonstrate that for smooth interfaces (Figure 4.44(a)), the analytical model is quite conservative for both typical concrete-to-concrete and concrete-to-UHPC 0%/UHPFRC

Figure 4.44 Comparisons of analytical and experimental results for (a) smooth and (b) rough interfaces [21]

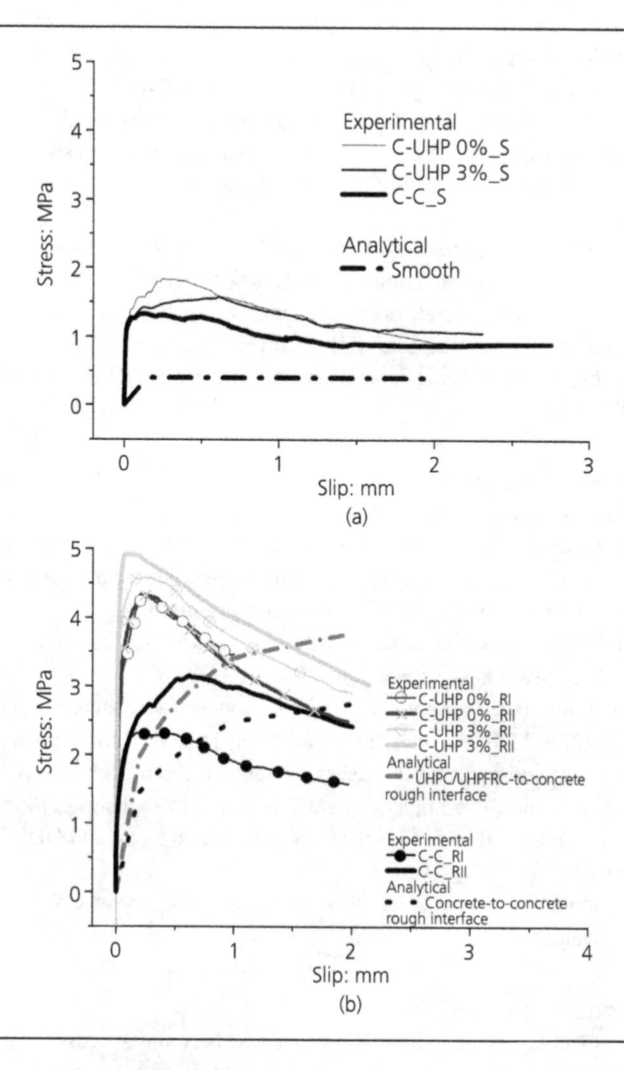

3% interfaces because the experimental values are nearly twice as high as the corresponding analytical results for high slip values (1–2 mm). Furthermore, the findings of the experimental investigation demonstrate that the highest shear stress values occur at very low slip values. For conventional concrete-to-concrete interfaces (C-C_RI, C-C_RII), the peak analytical stress values are in good agreement with the experimental data; however, for interfaces between concrete and UHPC (C-UHP 0%_RI, C-UHP 0%_RII) and concrete and UHPFRC (C-UHP 3%_RI, C-UHP 3%_RII), the experimental shear stress values are higher than the numerical data.

More specifically, the peak stress values for the concrete-to-UHPC 0% interfaces were found to be almost 14% higher than the corresponding analytical results, and the experimental results for these specimens exhibit relatively similar behaviour for the different degrees of roughening (C-UHP 0%_RI, C-UHP 0%_RII). The experimental peak stress values for the concrete-to-UHPFRC 3% interfaces (C-UHP 3%_RI, C-UHP 3%_RII) were shown to be greater than the analytical results. Peak stress for specimens C-UHP 3%_RI was 21% higher than the corresponding analytical value, while the peak stress for specimens C-UHP 3%_RII was found to be 30% higher than the corresponding analytical values. In contrast to the analytical results, which show the highest stress at the ultimate slip, the peak experimental shear stress values for rough interfaces were achieved at rather low slip values.

Concluding remarks The interfaces between UHPC/UHPFRC and conventional concrete were studied in this investigation. Triplet specimens with UHPFRC (3% steel fibre volume fractions) and conventional concrete were examined through push-off tests, while conventional concrete-to-concrete interfaces as well as corresponding triplet specimens with UHPC (without steel fibres) to conventional concrete interfaces were evaluated. The following results were found in this study.

- In comparison with the specimens with roughened interfaces, all the examined specimens with a smooth contact exhibit much lower shear stress values than expected.
- Shear stress values for the conventional concrete-to-concrete samples with rough (RI) and very rough (RII) interfaces were noticeably lower than for the corresponding UHPC 0%-to-concrete and UHPFRC 3%-to-concrete samples.
- The UHPFRC 3% to concrete surfaces with rough and very rough interfaces had the highest interface shear stress values.
- For the smooth interfaces for both conventional concrete-to-concrete and concrete-to-UHPC 0%/UHPFRC 3%, the analytical model was relatively conservative.
- For typical concrete-to-concrete interfaces, the experimental analytical results for the peak shear stress values were in agreement. The experimental shear stress values are higher than the numerical ones for interfaces between concrete and UHPC 0% and concrete and UHPFRC 3%
- For all the examined specimens, the peak experimental shear stress values occurred at very low slip values.

4.5.1.3 Size effect of UHFRC [30]

The use of UHPFRC layers or jackets has proved to be quite successful in strengthening applications and this is an area that has become increasingly popular over the past few years. However, there is a need for reliable numerical models to permit the widespread usage of UHPFRC in strengthening applications [30].

To determine the UHPFRC tensile stress–strain models, it is well-established practice to perform either direct tensile or flexural tests. These features are closely related to mix design, and various published research articles have looked at the impact of mix design, aggregate type and size and, more importantly, the dosage and properties of steel fibres [20, 22, 23, 27, 30–47]. Because the strain-hardening properties, which are directly related to the bridging effect of the fibres, are attributed to the bond between the fibres and the matrix, the behaviour of UHPFRC is typically significantly affected by the microstructure of the cementitious matrix and the characteristics of the fibres. Numerous researchers [34–40] have looked into how different fibre types, lengths and volume fractions affect the mechanical properties of UHPFRC. According to Paschalis and Lampropoulos [34], increasing the steel fibre content from 1% to 6% by volume increased both the compressive and tensile strengths by 72% and 92%, respectively. In their study of various fibre types, Hannawi *et al.* [35] discovered that the impact of the fibres on the compressive strength and elastic modulus of UHPFRC specimens was minimal at a volume proportion of 1%. The effects of steel fibre length and volume fraction on the mechanical characteristics and durability of the UHPFRC were investigated by Abbas *et al.* [36]. According to this study [36], the insertion of steel fibres greatly improved the tensile and flexural strength while only slightly improving the compressive strength. Additionally, it was noted that the fibres changed the failure pattern from brittle and explosive to ductile behaviour. The length of the fibres significantly affected the peak load-carrying capacity and load–deflection behaviour but had little impact on compressive strength. Microglass, hooked-steel, and microsteel fibres with a volume percentage of up to 2% were examined by Gesoglu *et al.* [37] and the mechanical properties were experimentally evaluated. They found that, regardless of the fibre type, a higher fibre content increased the compressive, tensile and flexural strength of the UHPFRC as well as its modulus of elasticity [37]. The mechanical characteristics of UHPFRC with up to 5% volume fraction of smooth steel fibres were studied by Kazemi and Lubell [38]. The main conclusions were that flexural and shear strength increased significantly as the steel fibre fraction increased. Wu *et al.* [39, 40] looked at how the fibre–matrix binding characteristics, as well as the compressive and flexural properties of UHPFRC, were affected by straight, corrugated and hooked fibres. The main conclusion of this study was that as fibre content and concrete age were increased, the compressive and ultimate flexural strengths were improved. Also, the addition of hooked fibres had significantly better pull-out bond strength and toughness than straight and corrugated fibres. Four-point bending experiments on UHPFRC beams with smooth steel fibres of various lengths were performed by Yoo *et al.* [41]. According to this study [41], fibre length, which improved fibre bridging capacity, greatly increased the load and toughness of the beams after the limit of proportionality. Additionally, microcracks were more prevalent in beams with longer fibre lengths.

The orientation of the fibres, which is influenced by the pouring technique and the size of the investigated specimens, is another crucial factor. The addition of fibres to the concrete matrix can significantly enhance the composite's overall mechanical performance, fracture behaviour and strength in tension, shear and flexure [34–40]. However, it can be quite challenging to obtain a uniform distribution of fibres throughout the mixture, particularly when a lot of fibres are being employed. In contrast to a proper and even distribution of fibres, which can ensure significantly enhanced mechanical properties, failure to achieve this goal may lead to reduced and even poor mechanical characteristics. This is more profound in case of UHPFRC, where (*a*) the interfacial bond between the fibres and the matrix is especially strong owing to the material's dense structure and (*b*) unreinforced matrices are exceedingly brittle owing to the lack of coarse aggregates. UHPFRC mechanical and fracture characteristics are heavily reliant on the uniform distribution of fibres. Any areas with a small percentage or with no fibres are possible weak points. The distribution of fibres in the mixture is influenced by a variety of variables, including the method used to incorporate the fibres, the frequency of vibration during compaction and the specimens' size and shape [31–33]. A few research studies on the so-called 'size effect' of UHPFRC [38, 42, 43] demonstrate the significance of specimen size for both compressive and flexural strength properties. Smaller samples often exhibit higher compressive and direct shear strengths, according to Kazemi and Lubell [38]. In contrast to thicker specimens (for example, those measuring 100 mm), specimens with a relatively small thickness (for example, those measuring 5 mm to 30 mm) can easily obtain a uniform distribution of fibres [22, 31, 32].

Mahmud *et al.* [42] looked into how the size of UHPFRC beams affected their flexural strength through three-point bending tests. The compressive strength of UHPFRC cubes of various sizes was studied by An *et al.* [43], who discovered that the larger specimens exhibited reduced compressive strength when compared with the smaller ones.

A sensitivity analysis was carried out by Awinda *et al.* [44] on UHPFRC prisms with different geometries as part of their experimental and numerical research. According to this study [44], for specimens with depths of 50 mm or less, the orientations and alignment of the fibres appear to be highly noticeable. Additionally, it was noted that more research is needed to examine the impact of variables such as fibre content and length on the numerical modelling of UHPFRC [44].

Since UHPFRC elements of very small thickness, including bridge decks and strengthening layers, have been applied widely in recent years, this is a particularly crucial aspect. It was demonstrated that adding extra UHPFRC layers or jackets can significantly increase the flexural and shear strength of RC constructions [20, 23, 27, 46]. The punching shear resistance of existing RC slabs has also been found to be significantly improved by the application of UHPFRC layers [47], and an analytical model can be used to calculate the improvement in punching shear of the composite/strengthened elements [48].

The UHPFRC strengthening layers that are used for the structural upgrade of existing structures are often of small thickness, having a geometry very different from that of the specimens used for the material characterisation. The tensile characteristics are derived either from prisms tested under flexural stress or from the direct tensile testing of dog-bone-shaped specimens with varying geometries in all of these applications; however, the size effect is normally not taken into consideration. In the case of UHPFRC specimens, the difference in the geometry of the examined specimens has a quite significant effect on the mechanical properties due to the orientation of the fibres, which is affected by the dimensions and, in particular, by the thickness of the structural elements.

The 'size effect' must be considered for the development of an improved simulation approach because the material properties of UHPFRC are highly affected by the dimensions of the specimens. A constitutive model for the tensile behaviour of UHPFRC was proposed in the literature using direct tensile tests and taking into account the size of the finite elements [30]. Experimental data on prisms of various geometries and thicknesses were used to evaluate the flexural and direct tensile characteristics of the UHPFRC specimens and these results were also used for the validation of the proposed approach [30].

This section focuses on the ways in which the volume fraction of steel fibres and the size of the examined specimens affect the flexural strength properties of UHPFRC and on a proposed appropriate method for the numerical modelling of UHPFRC elements with various geometries and thicknesses [30].

Description of the flexural test prisms and experimental results The findings of two different experimental investigations were collected and the results for UHPFRC prisms of various depths were used. Prisms of various geometries were subjected to flexural tests, and the findings were utilised to validate the proposed numerical model. In this section, information is provided about the experimental programmes including the concrete mix designs, the manufacturing process and the experimental results.

Two alternative UHPFRC mix designs, namely, UHPFRC-1 and UHPFRC-2, were examined in this investigation [30]. These were selected as typical mix designs with 3% and 6% fibre volume fractions. Nicolaides [31] describes the development of UHPFRC-1 at Cardiff University, whereas the experimental work of Hassan *et al.* [45] served as the base for UHPFRC-2. Table 4.10 presents the mix proportions for the two mixtures.

Both mix compositions have high-strength cement (52.5), microsilica and low water to binder ratios. The addition of superplasticisers was necessary to achieve the required workability. A maximum particle sand size of 0.5 mm was used for the UHPFRC-2 mix while the respective maximum size for UHPFRC-1 was equal to 0.6 mm. A substantial amount of ground granulated blast furnace slag (GGBS) was used in the UHPFRC-2 mix composition. Each composition also contained substantial amounts of steel fibres; specifically, 6% (468 kg/m^3) and 3% per volume (234 kg/m^3) were used for mixtures UHPFRC-1 and UHPFRC-2, respectively.

For UHPFRC-1, shorter (6 mm) and longer (13 mm) brass-coated steel fibres were combined, whereas in order to create UHPFRC-2, only one length of fibres (13 mm) was added to the mixture. All of the fibres had a 0.16 mm diameter, 3000 MPa tensile strength and 200 GPa modulus of elasticity.

Table 4.10 UHPFRC-1 and UHPFRC-2 mix compositions [30]

Material	Mix proportions: kg/m^3	
	UHPFRC-1	UHPFRC-2
Cement (52.5)	855	657
GGBS		418
Silica fume	214	119
Silica sand	940	1051
Superplasticisers	28	59
Water	188	185
Steel fibres	468	234

Both UHPFRC types were cast in the labs using the dry mixing method, which involves combining the dry ingredients (sand, silica fume, cement and GGBS) first before including any liquid ingredients. Steel fibres were also incorporated into the dry mixture in UHPFRC-1 immediately before the addition of water and superplasticiser. In UHPFRC-2, on the other hand, the steel fibres were added into the mixture soon after the liquid components. Both materials were produced using high-shear pan mixers.

The UHPFRC-1 moulds were cured at ambient conditions for 24 hours before the demoulded samples were put into a hot curing tank with water that was kept at a temperature of 90 °C. For nine days, the specimens were hot cured in the tank. To prevent any damage due to the heat curing, the temperature of the curing tank was progressively raised from 20 °C to 90 °C on the first day and then, on the ninth day, the temperature was progressively reduced from 90 °C to 20 °C. The heat curing was used to enable the full development of the mechanical properties at a reduced curing period. The UHPFRC-2 was also heat cured, with the specimens being placed in a hot curing tank at 90 °C after demoulding and tested after 14 days.

Two different layer depths from mix UHPFRC-1 (that is, 35 mm and 100 mm) were examined, whereas four different layer depths from mix UHPFRC-2 (that is, 25 mm, 50 mm, 75 mm and 100 mm) were studied. For each layer depth, at least three beams were prepared. To measure the direct tensile strength of UHPFRC-2, six dog-bone specimens were also produced and tested. Benson and Karihaloo previously conducted a study to estimate the equivalent value of the direct tensile strength of UHPFRC-1 [49]. Typical photographs from the preparation of beams with various layer thicknesses are shown in Figure 4.45.

The results of the flexural testing of the examined specimens with various section depths are presented in this section. For each of the examined depths, at least three specimens were tested. The experimental set-ups for (*a*) the three-point bending test (UHPFRC-1) and (*b*) the

Figure 4.45 Prism preparation of specimens with varying depths [30]

(a) (b) (c)

four-point bending test (UHPFRC-2) are schematically shown in Figure 4.46, along with the measurements of the overall and testing spans and the loading conditions.

UHPFRC-1 prisms were tested in three-point bending under displacement control with the application of a load to the middle of the span. For each beam, two measurements were recorded: the applied force and the mid-span deflection. A single LVDT was positioned at the centre of the testing beam to measure the vertical deflection (Figure 4.47(a)).

Figure 4.46 Illustration of the loading and support conditions of (a) the bending test used for UHPFRC-1 and (b) the bending test of UHPFRC-2 (dimensions in mm) [30]

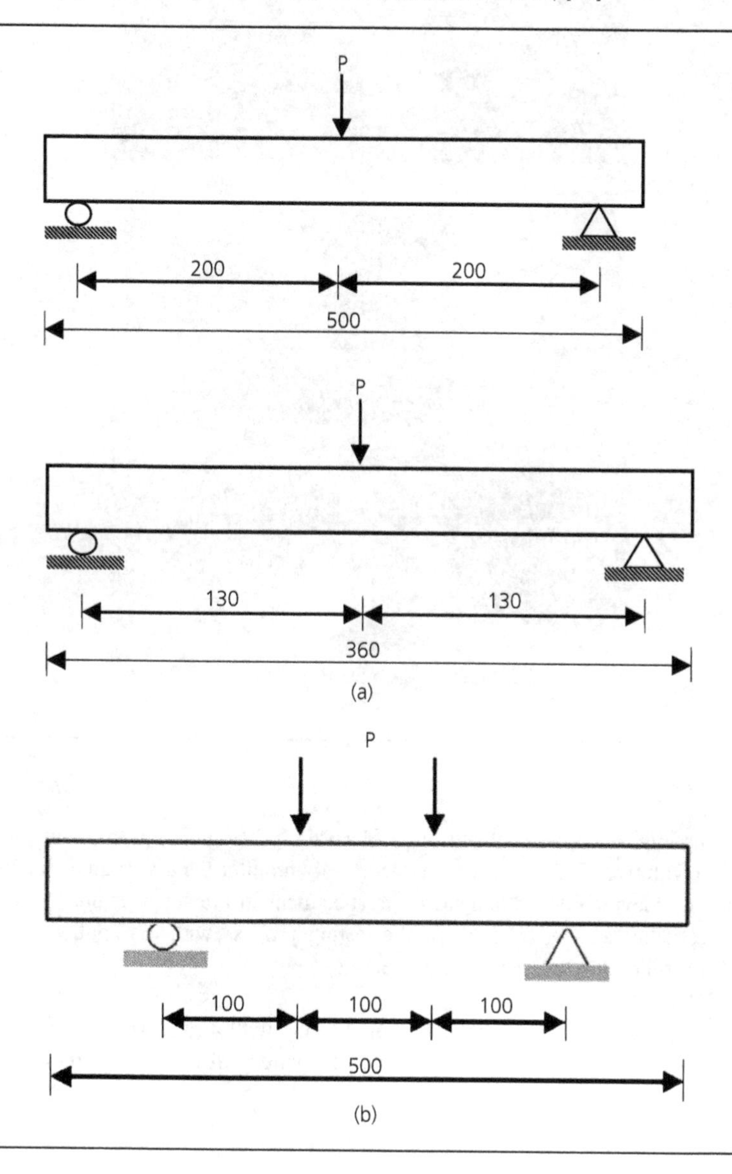

Figure 4.47 Testing set-up for (a) UHPFRC-1 and (b) UHPFRC-2 specimens [30]

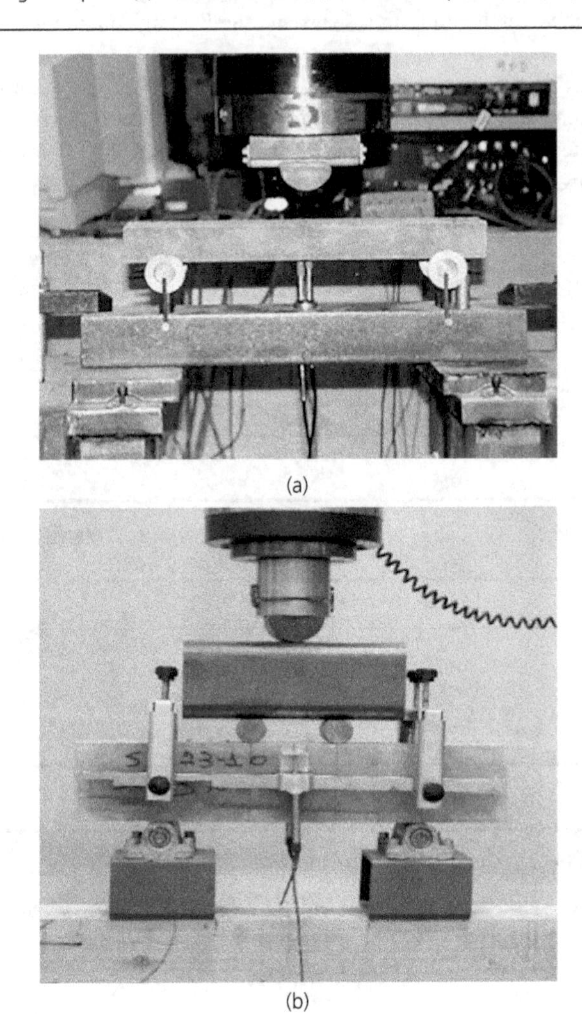

(a)

(b)

Four-point loading was used for the testing of UHPFRC-2 (Figure 4.47(b)). An external steel frame (yoke) with two LVDTs was used to accurately monitor the mid-span deflection, which eliminated any unavoidable settlement of the specimens at the supports during the loading. JSCE [50] specifications were used for the testing process with the application of a displacement controlled load at a rate of 0.001 mm/s.

All of the tested samples show a similar failure mode with a primary crack forming at the middle of the span. Figure 4.48 displays the crack patterns for a few common prisms with various depths.

Figure 4.48 Typical failure modes for prisms with (a) 25 mm, (b) 50 mm, (c) 75 mm and (d) 100 mm depths [30]

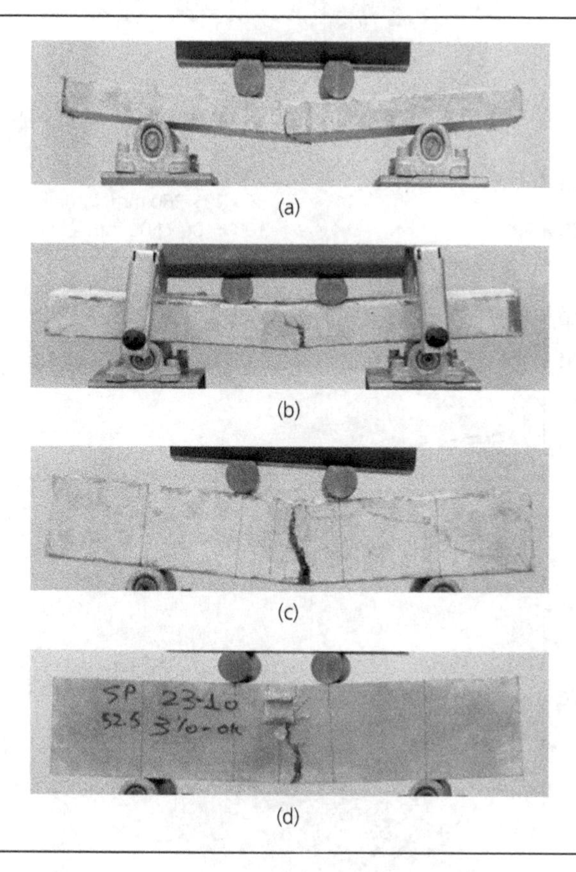

Figures 4.49(a) and 4.49(b) show the load–deflection results for all the specimens tested for UHPFRC-1 and UHPFRC-2, respectively [30].

For both UHPFRC-1 and UHPFRC-2, the flexural strength was calculated using the values of Figure 4.49.

Equation 4.6 was used for the calculation of the flexural strength σ_t:

$$\sigma_t = \frac{M \times y}{I} \tag{4.6}$$

where:

M is the bending moment
I is the moment of inertia
y is the distance of the centroid from the extreme fibre.

Figure 4.49 Experimental load–deflection results for all the examined specimens for (a) UHPFRC-1 and (b) UHPFRC-2 [30]

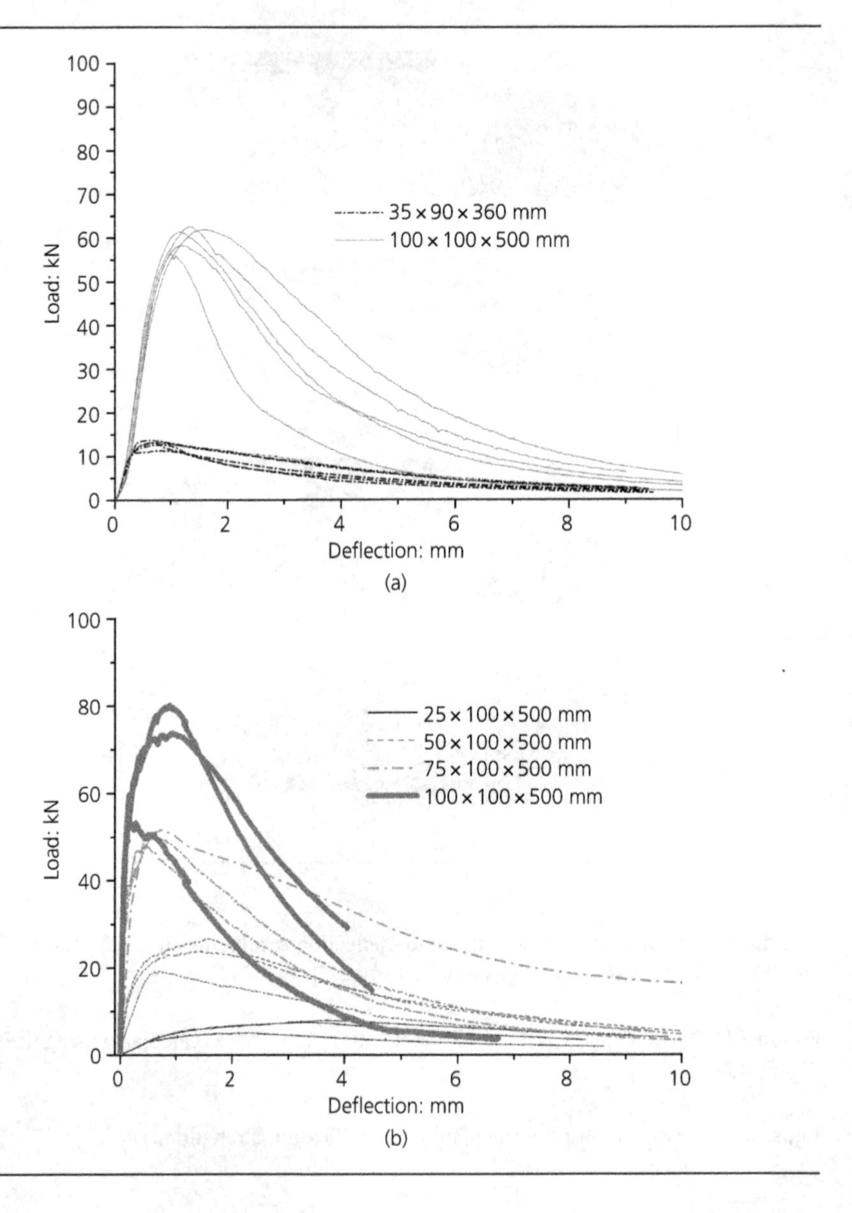

The following models are derived for the three-point (Equation 4.7) and four-point (Equation 4.8) bending tests of UHPFRC-1 and UHPFRC-2 using Equation (4.6).

$$\sigma_{t\text{-}3p} = \frac{3 \times P \times L}{2 \times b \times d^2} \tag{4.7}$$

$$\sigma_{t-4p} = \frac{P \times L}{b \times d^2} \tag{4.8}$$

where:

$\sigma_{t\text{-}3p}$ and $\sigma_{t\text{-}4p}$ are the flexural strength values calculated from the three-point and four-point bending tests (MPa)
P is the peak load (N)
L is the effective span length (mm)
b is the width of specimen (mm)
d is the depth of the specimen (mm).

The average calculated flexural strength results along with the scatter plot are shown in Figure 4.50 for each of the various examined thicknesses for UHPFRC-1 and UHPFRC-2.

Based on the results shown in Figure 4.50, the so-called 'size effect' is confirmed through the flexural strength reduction with the increment of the section depth of the specimens. In comparison with specimens having smaller thicknesses, where there is a more even distribution of the fibres, and therefore enhanced flexural strength is achieved, thicker specimens (for example, 100 mm) show significantly reduced flexural strength values, which is attributed to the uneven distribution of fibres in the parts of the specimens. The larger percentage of steel fibres in UHPFRC-1 also results in a higher total flexural strength than the corresponding values in UHPFRC-2, as expected. However, the results of Figure 4.50 show a somewhat faster rate of reduction for UHPFRC-1 owing to its larger proportion of steel fibres (UHPFRC-1 contains 6% steel fibres compared with UHPFRC-2, which has 3% steel fibres).

Figure 4.50 Results of flexural strength against prism depth for UHPFRC-1 and UHPFRC-2 [30]

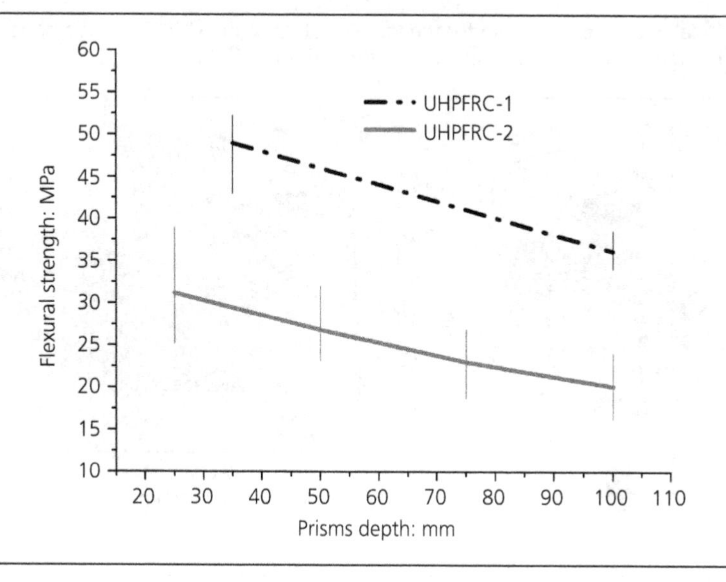

Proposed constitutive models for numerical analyses Direct tensile tests were conducted to calculate the tensile stress characteristics of both UHPFRC-1 and UHPFRC-2. These results were used to determine the UHPFRC constitutive models. For the constitutive modelling of UHPFRC, direct tensile strength and Young's modulus were used. Six UHPFRC-2 dog-bone samples were tested in direct tension [23]. The tensile loading was applied using two steel grips, and all specimens were evaluated with displacement control at a rate of 0.007 mm/s. The configuration shown in Figure 4.51(a) was used to compute the extension and the corresponding strain values which were measured using LVDTs. Figure 4.51(b) presents the results of tensile stress against strain for the UHPFRC-2 specimens. In this figure, the distribution of stress and total strain for each individual specimen together with the average results are illustrated [23]. The results of Figure 4.51(b) show that UHPFRC-2 has a tensile strength that ranges from 11.74 MPa to 14.20 MPa, while the average strength value was calculated to be equal to 12.15 MPa.

The constitutive model shown in Figure 4.52 reflects the stress–strain distribution following the end of the elastic section and was developed using the experimental results shown in Figure 4.51(b). This model has a linear portion that extends up to the highest tensile stress value (f_t), followed by a bilinear branch that descends (Figure 4.52). To assess the accuracy of the model, this will be utilised to numerically simulate both UHPFRC-1 and UHPFRC-2.

The numerical simulations were performed using the ATENA program, and the model shown in Figure 4.52 was used to describe the tensile stress–strain behaviour using experimental values for the tensile strength and Young's modulus. Experimental testing on UHPFRC-1 resulted in a tensile strength of 16MPa, while Young's modulus and compressive strength were equal to 48 GPa and 193.6 MPa, respectively [31]. A tensile strength of 11.5 MPa and Young's modulus of 57.5 GPa were found for UHPFRC-2, and compressive strength was calculated to be 164 MPa.

Figure 4.51 (a) Experimental set-up for the tensile testing of the UHPFRC-2 dog-bone specimens and (b) stress versus strain experimental results for UHPFRC-2 [23]

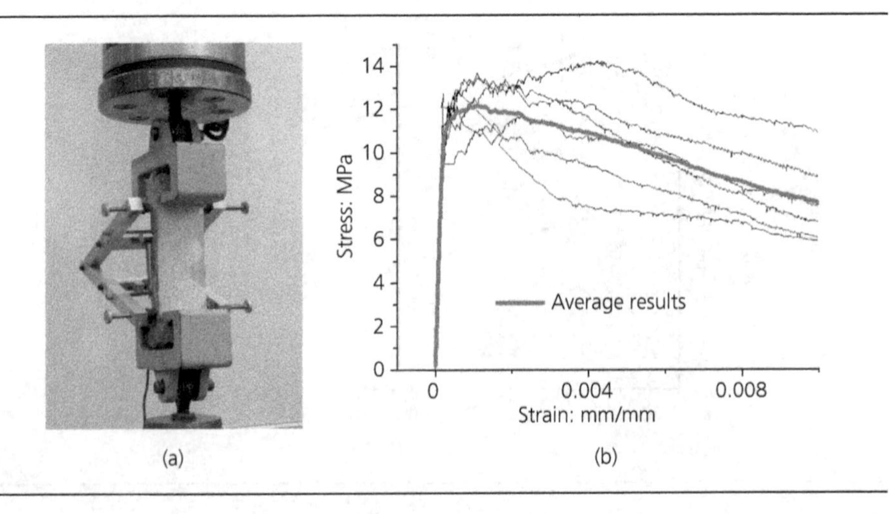

Figure 4.52 UHPFRC constitutive model [30]

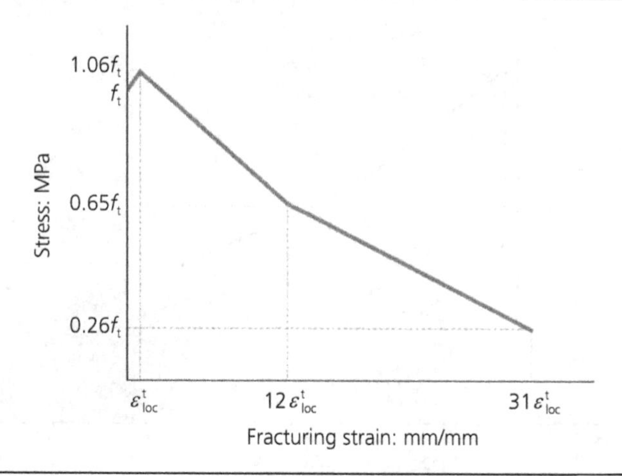

With characteristic size (l_{ch}) equal to 2 mm and finite element size (l_t) equal to 65 mm, the strain values for the characteristic points of the model in Figure 4.52 were calculated using strain ε^t_{loc} equal to 0.042 [23]. The reliability of this model was found to be highly dependent on the values of the characteristic size and the mesh size of the elements of the numerical models, which have a significant impact on the results in the post-crack region. This model was found to be able to accurately predict the behaviour of UHPFRC-2.

It is crucial to create models that can reliably predict UHPFRC behaviour regardless of the size of the finite element models.

The construction of a model that takes into consideration the size of the elements and can be used to precisely anticipate the behaviour of various geometries of UHPFRC specimens is the main topic of this part of the study. According to the model shown in Figure 4.53, the proposed tensile stress–strain characteristics are defined each time based on the size of the finite elements (l_t).

Both UHPFRC-1 and UHPFRC-2 were examined and the numerical models are illustrated in Figures 4.54 and 4.55.

The finite element size ($l_{t\ new}$) for the specimens of Figures 4.55(b)–4.55(d) was 14.7 mm, whereas the corresponding value for the specimen of Figure 4.55(a) was 15 mm. The UHPFRC constitutive model was modified using the method shown in Figure 4.53 by changing the characteristic size value ($l_{ch} = 2$ mm), which was initially proposed for $l_t = 65$ mm by multiplying it by $l_{t\ new}/l_t$ (that is, for $l_{t\ new} = 14.7$ mm, $l_{t\ new}/l_t = 0.2$ and $l_{ch\ new} = 0.2 \times l_{ch} =$ 0.4 mm).

Figure 4.53 UHPFRC mesh-size-dependent constitutive modelling [30]

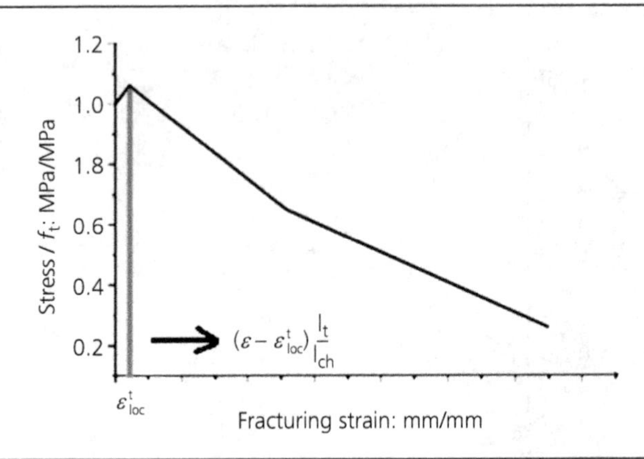

Figure 4.54 UHPFRC-1 numerical models for (a) 25 × 90 × 360 mm and (b) 100 × 100 × 500 mm prisms [30]

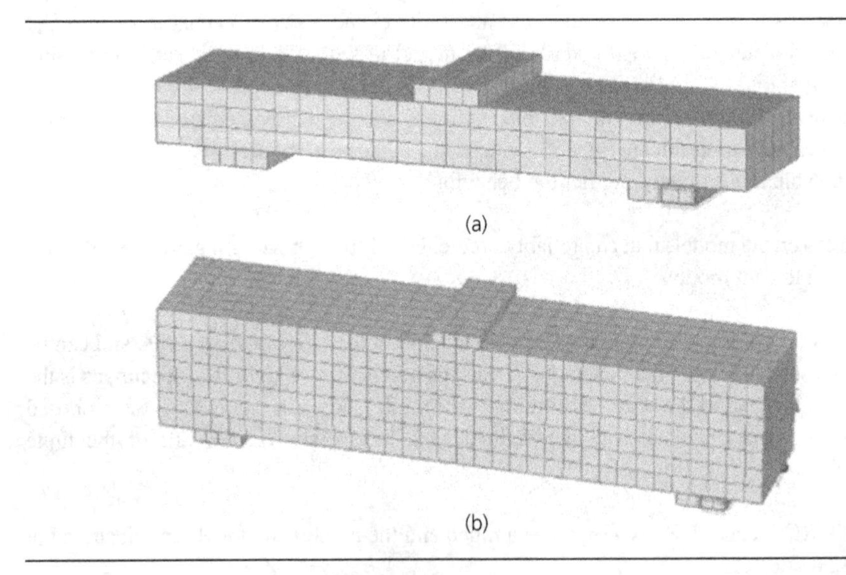

To replicate the conditions of the experimental testing, simply supported conditions were applied to all of the studied specimens, and a monotonically increasing displacement was applied to the middle of the span. In the following section, comparisons of the numerical and analytical results are shown.

The comparison of the numerical finite element analysis (FEA) results with the experimental results are presented in Figure 4.56 for UHPFRC-1, while Figure 4.57 shows the corresponding results for UHPFRC-2. The results of Figures 4.56 and 4.57 demonstrate that,

Figure 4.55 UHPFRC-2 numerical models for (a) 25 × 100 × 500 mm, (b) 50 × 100 × 500 mm, (c) 75 × 100 × 500 mm and (d) 100 × 100 × 500 mm prisms [30]

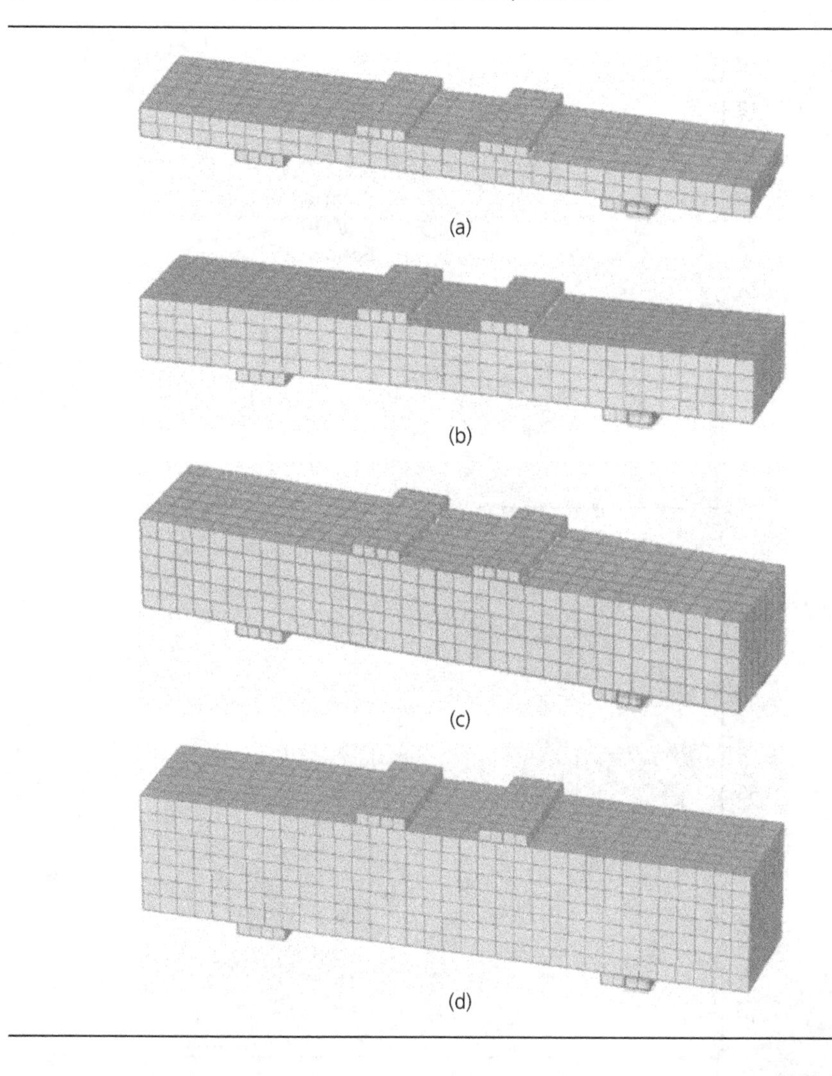

for both UHPFRC-1 and UHPFRC-2, the numerical modelling results are in good agreement with all the experimental findings despite the variation of the prism dimensions.

More specifically, the results of this investigation show that, in the majority of the studied samples, initial stiffness, maximum stress and post-cracking behaviour could all be accurately predicted using the proposed numerical approach. The experimental results show a quite significant scatter in the results of the multiple samples of each mix and geometry, which is attributed to the random orientation and distribution of the fibres. The numerical results are generally in close agreement with the experimental findings.

Figure 4.56 Comparison of experimental with numerical results for UHPFRC-1 specimens with (a) 25 mm and (b) 100 mm section depths [30]

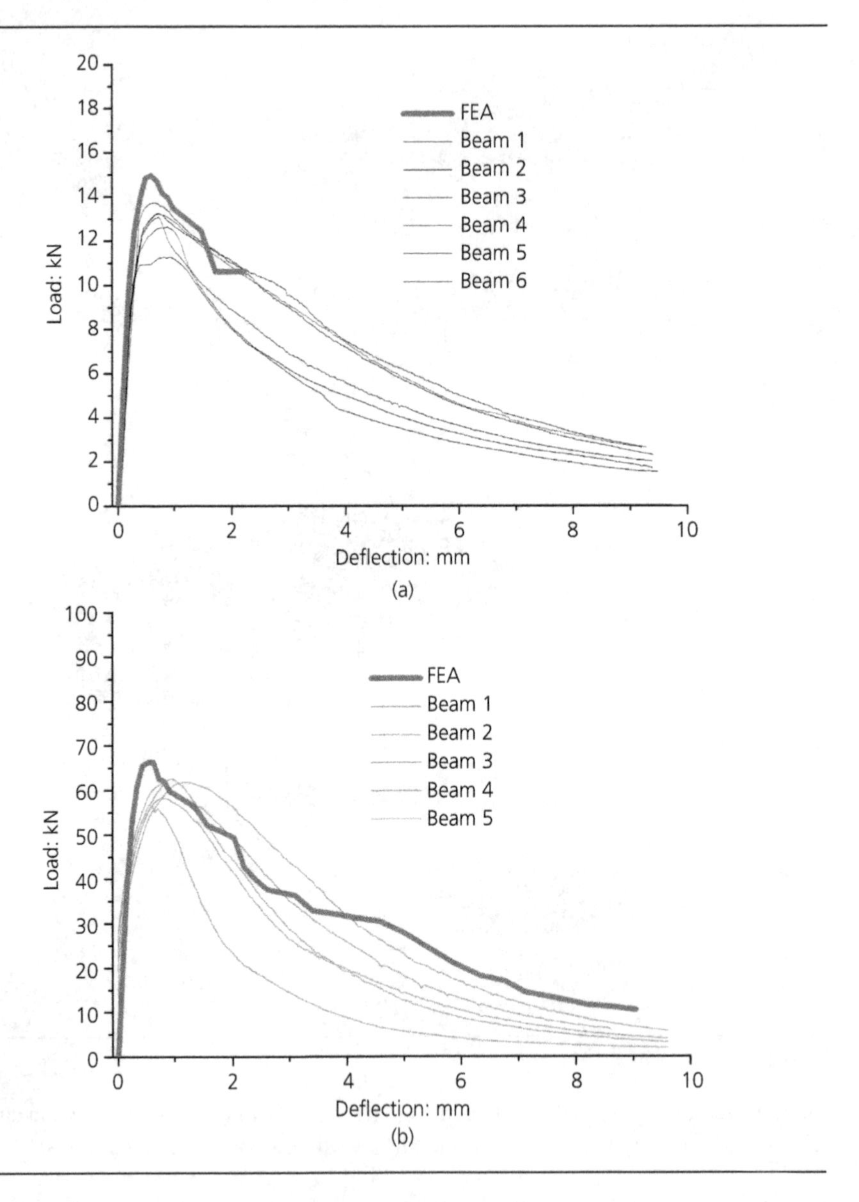

It is also important to note that the numerical results for specimens with relatively shallow depths (that is, 25 mm, 50 mm and 75 mm) are in most cases in good agreement with the average of the experimental results, whereas the numerical results for specimens with 100 mm are marginally superior to the highest values of the experimental results. This is because of the non-uniform distribution and orientation of the fibres, which is more significant in the case of specimens with large thickness (that is, 100 mm).

Figure 4.57 Comparison of experimental with numerical results for UHPFRC-2 specimens with (a) 25 mm, (b) 50 mm, (c) 75 mm and (d) 100 mm section depths [30]

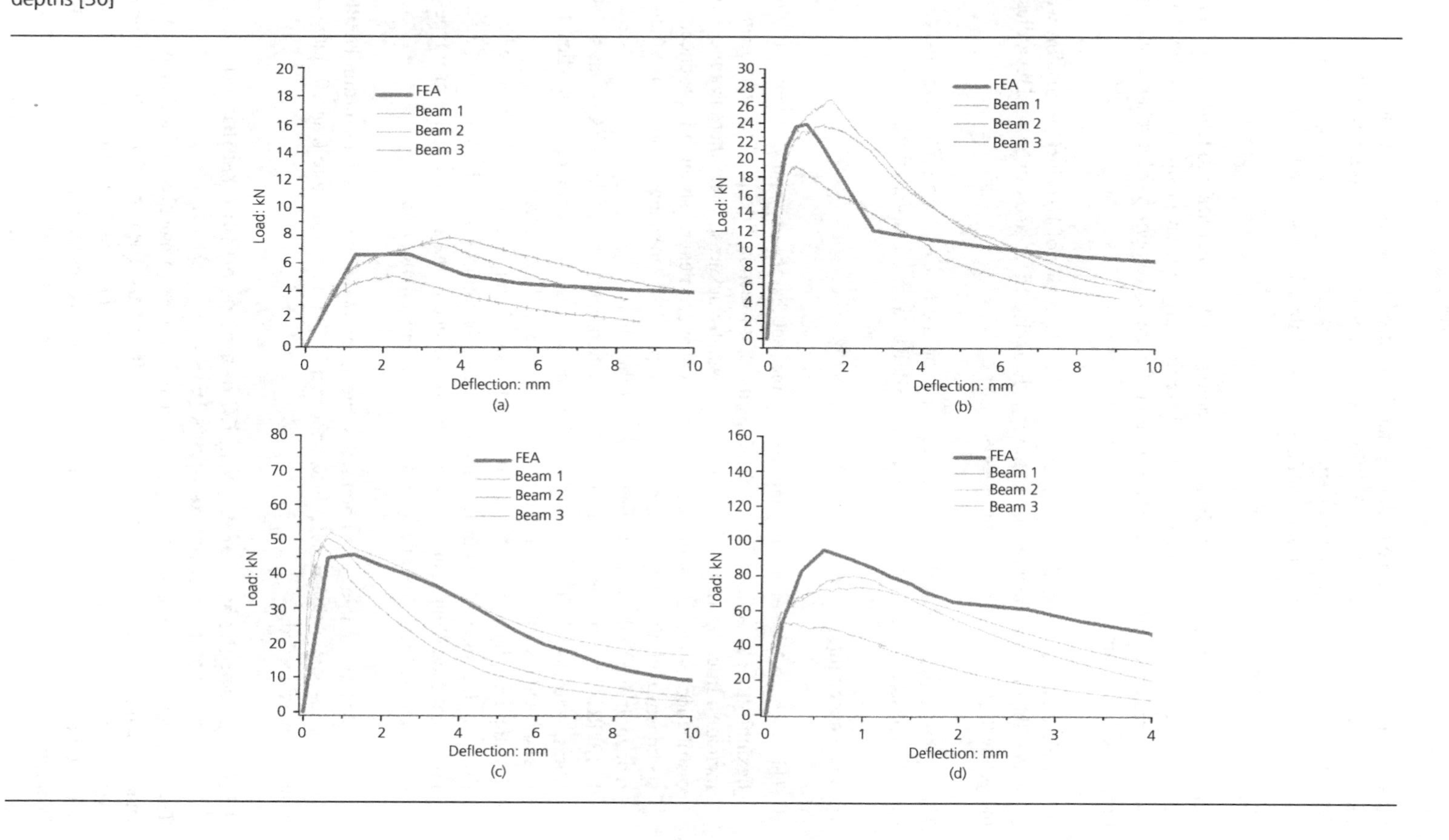

As a result, the suggested methodology can be utilised to precisely simulate the structural performance of the relatively thin UHPFRC layers which have thicknesses normally no greater than 100 mm. These findings demonstrate how the suggested approach may be used to simulate UHPFRC specimens of various thicknesses while minimising the influence of the so-called 'size effect'.

Concluding remarks In this study, a methodology for the numerical modelling of UHPFRC was suggested. The results of extensive experimental work on prisms with various fibre volume percentages and geometries were used for the development of a reliable numerical procedure for the simulation of their structural behaviour.

A suitable constitutive model is proposed for the behaviour of UHPFRC in tension which eliminates the effect of the influence of the so-called 'size effect' and can be used regardless of the geometry of the finite element models.

The suggested model, which takes into account the mesh size of the finite elements, can be used to precisely simulate the behaviour of UHPFRC layer of varying thicknesses.

The main concluding remarks are summarised below.

- The so-called 'size effect' is confirmed by the fact that the experimental values of the flexural strength of the prisms reduces as the cross-sectional depth of the specimens increases. In comparison with specimens of small thickness, where there is a more even distribution of the fibres and, consequently, enhanced flexural strength is produced, the specimens with increased thickness (for example, 100 mm) show a significant strength reduction due to the uneven distribution of the fibres.
- UHPFRC-1 specimens with a larger percentage of steel fibres (UHPFRC-1 has 6% steel fibres, whereas UHPFRC-2 has 3%), show an increased reduction rate of the flexural strength values with the increment of the thickness of the specimens.
- The applicability of the proposed method for simulating UHPFRC specimens with different dimensions, eliminating the size effect, is confirmed by the ability of the numerical modelling approach to accurately predict all the examined specimen types (with different fibre volume fractions and different geometries).

The suggested methodology can be used to precisely model the behaviour of thin UHPFRC layers. For the simulation of UHPFRC specimens with thickness more than 100 mm, additional study is needed.

4.5.2 Strengthening of existing RC beams using fibre reinforced geopolymer concrete layers [51]

The application of additional RC layers is one of the most commonly used techniques for the enhancement of the flexural strength of existing RC beams. The connection between the new layer and the existing concrete in this scenario is a critical factor in determining the effectiveness of the technology under investigation, because insufficient connection at the interface may result in the strengthened elements failing earlier than expected. The corrosion of the reinforcement of the layers is another important factor for the durability of the strengthened elements.

In this section, the application of additional layers of fibre-reinforced geopolymer concrete reinforced with steel bars was studied, with the aim of offering enhanced structural performance in addition to sustainability and durability enhancements [51].

The addition of conventional RC layers or jackets to existing RC structures is one of the most commonly used reinforcing techniques, and it was demonstrated that this approach can significantly enhance the structural performance of existing elements [19, 52].

However, it is essential to make sure that enough connection is offered at the interface between the old and the new RC elements to prevent premature structural failures [19]. The resilience and corrosion resistance of the steel reinforcement are other key factors in the structural performance of the strengthened elements. When the concrete cover is insufficient, chloride ions can flow through the cover concrete and reach the steel bars, causing the passive film to be destroyed and the steel bars to corrode. This has a considerable impact on the structural performance of the RC elements [53–55].

In order to enhance the mechanical performance of conventional concrete, in particular its ductility and post-cracking performance, the addition of steel fibres to the concrete mix has been thoroughly studied over the past 20 years. As a result, high-performance cementitious materials such as UHPFRC have been developed. For the flexural and shear strengthening of pre-existing structural elements, the inclusion of UHPFRC layers was shown to be extremely effective [23, 46]. In Section 4.5.1, the application of UHPFRC layers for the strengthening of RC beams has been thoroughly presented.

In earlier research initiatives, the development of fibre-reinforced geopolymer concrete (FRGC) was investigated as a sustainable and environmentally friendly material with improved mechanical performance and strain-hardening properties (Section 3.3).

In this section, the use of FRGC layers reinforced with steel bars for the strengthening of existing RC beams is presented. For the evaluation of the effectiveness of the examined technique, strengthened RC beams were tested under flexural loading following accelerated corrosion tests using the induced current method. To assess the efficacy of the suggested method, the results of specimens with conventional RC layers and specimens without induced corrosion were also used.

4.5.2.1 Description of the experimental procedure and results

In this section, the geometry and dimensions of the investigated specimens are described, along with the material parameters and a description of the FRGC preparation.

In this study [51], a total of eight beams were investigated. Normal-strength conventional concrete (NSC) was used to strengthen four beams; two of the strengthened beams were examined after the effect of accelerated corrosion to the steel bars of the layers (NSC-S-corr), while the other two were not exposed to corrosion (NSC-S). Four further examples were enhanced using layers of polyvinyl-alcohol-fibre-reinforced geopolymer concrete (PVAFRGC); two of the layers were examined after the effect of accelerated corrosion to their steel bars (PVAFRGC-S-corr), while the other two were not exposed to corrosion (PVAFRGC-S).

Two 10 mm diameter bars (210) with a yield stress of 530 MPa were placed on the tensile side of the initial beams (Figure 4.58). In the shear span, stirrups with a diameter of 8 mm and a yield stress of 350 MPa were spaced 90 mm apart. The initial beams were cast using ordinary Portland cement, which had a mean cube compressive strength of 32 MPa on the testing day.

Three months after the initial beams were cast, the additional layer was added. With an air chipping hammer, the surfaces of the initial beams were roughened to a depth of 2–4 mm. The additional layer was 50 mm thick, and the layers were strengthened with two bars each measuring 10 mm in diameter and having a yield stress of 530 MPa (Figure 4.58).

Fly ash, ground granulated blast furnace slag (GGBS) and undensified silica fume were all employed for the PVAFRGC. Additionally, silica sand with a maximum particle size of 0.5 mm was included in this mixture. Potassium hydroxide and potassium silicate solution were used as activators. With regard to the fibres, 12 mm polyvinyl alcohol (PVA) fibres at 2% (by volume) were added to the mixture. A typical cement-based mix with predicted strength characteristics comparable to those of the PVAFRGC was employed for the NSC.

Table 4.11 presents the PVAFRGC and NSC mix compositions.

A Zyklos high-shear mixer (Pan Mixer ZZ 75 HE) was utilised for the material mixing. Sand, alkaline liquid and geopolymer binder (fly ash, GGBS, and silica fume) were added to the mixer in that order. Prior to mixing with the solid phase, the liquid phase was generated by combining potassium silicate solution with water and polycarboxylate-based superplasticiser for 5 minutes. After five minutes of dry mixing, the liquid phase was added, and the mixer continued to run for an additional five minutes.

To guarantee consistent fibre dispersion in the geopolymer mix, fibres were gradually added after being sieved through a suitable steel mesh at the top of the mixer. Sand was added last to the mixer, which ran for an additional three minutes to complete the 13-minute mixing process.

The samples were demoulded, coated in plastic sheets to control moisture loss, and allowed to cure at room temperature until the test date.

Standard compressive tests on cubes with 100 mm sides and direct tensile testing on dog-bone specimens with a cross-section of 13 mm by 50 mm were used to determine the mechanical properties of both NSC and PVAFRGC.

Figure 4.58 Geometry of the strengthened beams [51]

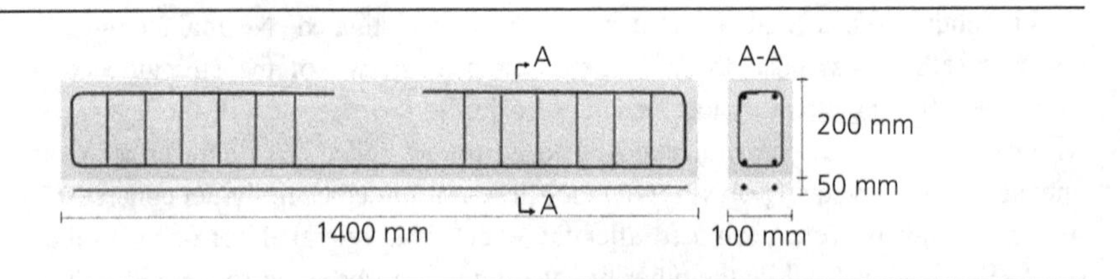

Table 4.11 PVAFRGC and NSC mix compositions [51]

Material	Mix proportions: kg/m^3	
PVAFRGC	NSC	
Fly ash	388	–
Slag	310	–
Silica fume	78	–
Cement	–	380
Alkaline activator	93	–
Water	194	194
Sand	1052	920
Gravel	–	800
PVA fibre	26	–

Based on the results of these tests, it was determined that the compressive and tensile strength of PVAFRGC were 46 MPa and 3.5 MPa, respectively, whereas the corresponding values for NSC were 43 MPa and 3 MPa, respectively. This confirms that similar material properties were attained for these two different materials used for the additional layers.

To simulate the impact of corrosion on the steel reinforcement of the additional layer, the approach presented in Section 3.3.6 was employed (Figure 3.20).

The bottom surface of the container was connected to the negative terminal of a DC power source, and a steady current of 300 mA was supplied for 30 days between the reinforcing bars (anode) of the extra layer and a copper mesh (cathode) (Figure 3.20).

Four-point bending tests were carried out to assess the structural performance of the strengthened beams, and the load against mid-span displacement was recorded. LVDTs were used to monitor the slip at the interface between the layers and the initial beams in addition to the load–deflection results (Figure 4.59).

All of the specimens studied had effective spans of 1200 mm, and they were tested under displacement control at a loading rate of 0.004 mm/s.

Following the completion of the four-point loading tests, the corrosion effect was assessed on bars of the corroded specimens by measuring the mass loss and correlating it with the experimental results of the bending tests.

Figure 4.60 shows the failure mechanism and fracture pattern during the four-point bending tests of the reinforced beams with reinforced NSC layers, while Figure 4.61 shows the respective failure modes for the strengthened beams with reinforced PVAFRGC layers.

Figure 4.59 Testing set-up for the flexural tests [51]

Figure 4.60 Illustration of mode of failure for (a) NSC-S-1, (b) NSC-S-2, (c) NSC-S-corr-1 and (d) NSC-S-corr-2 [51]

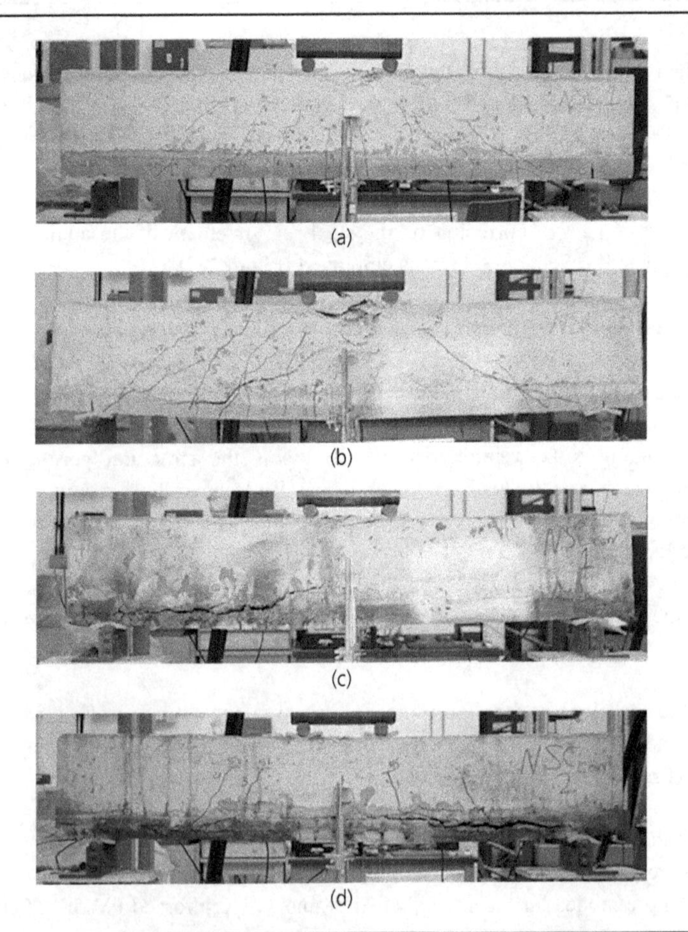

(a)

(b)

(c)

(d)

Figure 4.61 Illustration of mode of failure for (a) PVAFRGC-S-1, (b) PVAFRGC-S-2, (c) PVAFRGC-S-corr-1 and (d) PVAFRGC-S-corr-2 [51]

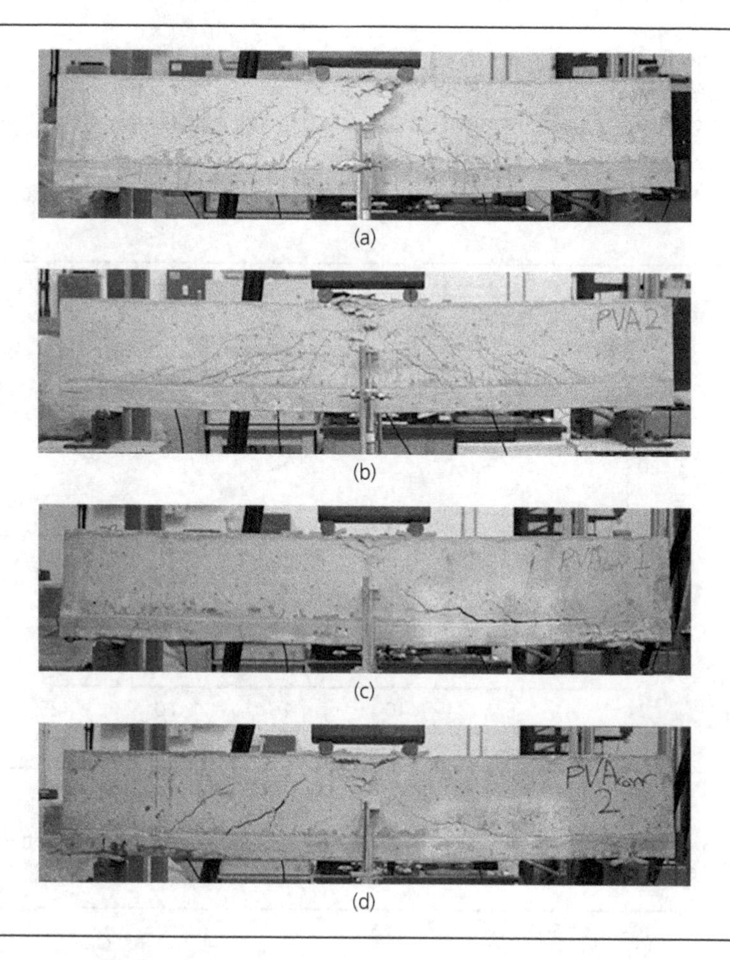

According to the failure modes shown in Figures 4.60 and 4.61, in the case of strengthened RC beams with NSC layers without corrosion, concrete crushing took place first followed by shear cracking, which continued at the interface. Peeling off of the concrete cover of the additional layer and shear cracks, which continued at the level of the interface, occurred in the case of the corroded specimens strengthened with NSC layers (Figures 4.60(c) and 4.60(d)), which is related to the significant effect of corrosion.

The specimens strengthened with PVAFRGC layers (Figure 3.61) had overall similar failure mode to the specimens strengthened with NSC layers, with the main difference being that, in the case of specimens strengthened with PVAFRGC layers, there was no separation of the concrete cover of the corroded specimens (Figures 4.61(c) and 4.61(d)), which confirms the improved resistance to corrosion of PVAFRGC. The majority of the specimens under investigation (Figures 4.60 and 4.61) experienced cracking at the interface at or very close to the maximum load value.

Figures 4.62 and 4.63 show the load–deflection findings for specimens strengthened with NSC and PVAFRGC layers, respectively.

Based on the findings of Figure 4.62, the average maximum load of the NSC-enhanced beams was determined to be 109.5 kN without corrosion, whereas the corresponding value for the corroded specimens was much lower and equivalent to 68.7 kN, which represents a 37% reduction.

Figure 4.62 Experimental load–deflection results of RC beams strengthened with additional NSC layers [51]

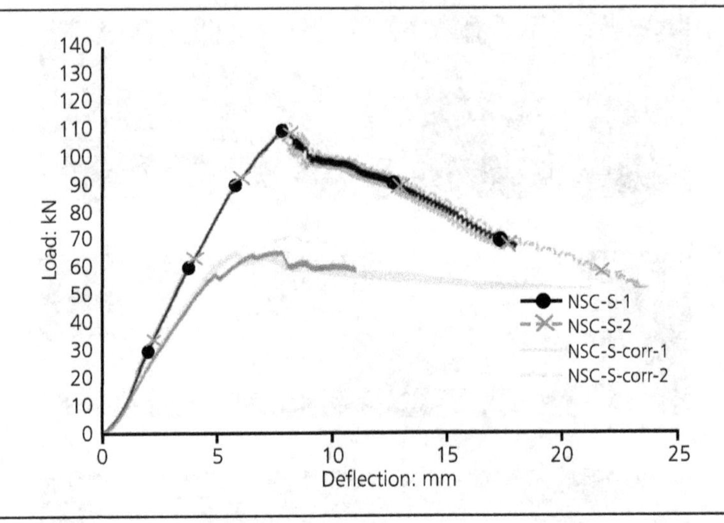

Figure 4.63 Experimental load–deflection results of RC beams strengthened with additional PVAFRGC layers [51]

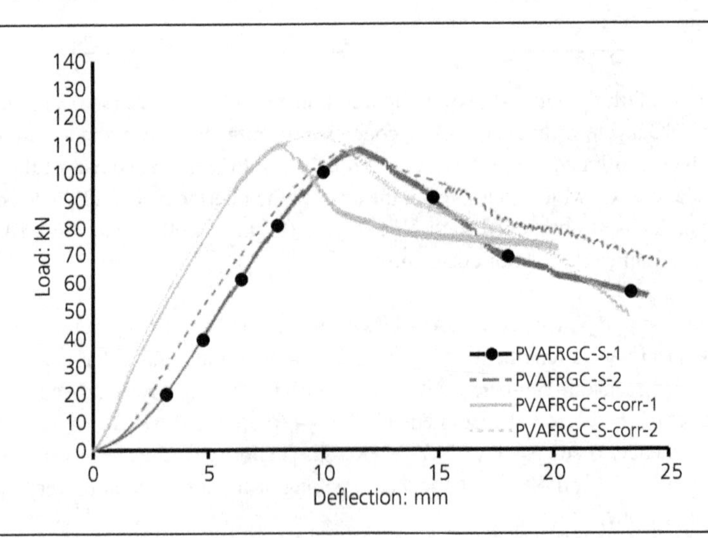

The average maximum load of the uncorroded specimens in the case of beams strengthened with PVAFRGC layers (Figure 4.63) was found to be 106.9 kN, but the corresponding value for the corroded specimens was found to be quite close and equal to 112.1 kN. This shows that corrosion had no impact on the maximum load value and confirms the positive impact of the PVAFRGC coating on the ability of the strengthened specimens to resist corrosion.

The reinforcing bars were mechanically cleaned with a strong metal brush to remove any adhering corrosion products, and the mass loss was measured in order to further assess the impact of the degree of corrosion. The mass loss for specimens strengthened with NSC layers was found to be 11%, whereas the reduction for specimens strengthened with PVAFRGC layers was found to be significantly lower and equal to 7.5%, which confirms the beneficial effect of PVAFRGC in protecting steel bars against corrosion.

In all of the examined specimens, the interface slip was recorded during the testing and these values were utilised to assess the interface properties.

The maximum interface slips at the ultimate load as well as the load values for slip equal to 0.2 mm, 0.8 mm and 1.5 mm are reported in Table 4.12 using the slip measurements of RC beams strengthened with NSC overlay. The Greek Code of Interventions [29] proposes slip values of 0.2 mm, 0.8 mm and 1.5 mm as the maximum slip values for immediate occupancy, life safety and collapse-prevention performance levels.

Based on the results of Table 4.12, it can be shown that specimens strengthened with PVAFRGC layers generally had lower slip values evaluated at the maximum load than specimens strengthened with NSC layers. For all four of the specimens strengthened with PVAFRGC layers, an average maximum slip value of 0.63 mm was measured at the maximum load, which is considerably less than the respective slip for the specimens strengthened with

Table 4.12 Selected load results for various characteristic interface slip values during the testing of the examined specimens [51]

Specimen	P: kN	S_{max}: mm	$P(s = 0.2$ mm): kN	$P(s = 0.8$ mm): kN	$P(s = 1.5$ mm): kN
NSC-S-1	106.8	1	60.0	87.0	–
NSC-S-2	112.2	0.69	60.7	–	–
NSC-S-corr-1	66.1	0.32	45.7	–	–
NSC-S-corr-2	71.3	1.33	36.8	65.7	–
PVAFRGC-S-1	109.0	0.70	57.9	109.0	–
PVAFRGC-S-2	104.7	0.41	67.4	–	–
PVAFRGC-S-corr-1	110.5	0.82	80.1	80.6	–
PVAFRGC-S-corr-2	113.6	0.6	79.3	–	–

NSC layers, which was determined to be equivalent to 0.84 mm. This suggests that improved interface conditions are obtained for specimens enhanced with PVAFRGC layers.

The Greek Code of Interventions [29] model for roughened interface (Equation 4.9) can be used to estimate the interface shear strength (τ_{fud}) of strengthened RC beams.

$$\tau_{\mathrm{fud}} = \begin{cases} 0.25 \times f_{\mathrm{ct}}, & \text{smooth interface} \\ 0.75 \times f_{\mathrm{ct}}, & \text{rough interface} \\ f_{\mathrm{ct}}, & \text{use of shotcrete} \end{cases} \tag{4.9}$$

where f_{ct} is the minimum value (between the old and new) of the concrete's tensile strength.

Additionally, using Equation 4.10, the appropriate interface shear stress (τ_x) can be calculated in accordance with British Standard BS 8110-1 [56], and the corresponding outcomes for the studied specimens are shown in Table 4.13.

$$\tau_x = \frac{V_{\mathrm{sd}}}{b \times z} \tag{4.10}$$

where:

V_{sd} is the shear force of the examined section of the beam
b is the width of the interface
z is the lever arm of the composite section.

Based on the results shown in Table 4.13, shear stress at maximum load (τ_x) was larger than shear strength (τ_{fud}) for all of the specimens under consideration, which is consistent with the experimental results that show high values of interface slips and cracking at the level of the interface during bending tests (Figures 4.60 and 4.61). Also, it should be mentioned that the reduced load capacity of the corroded specimens leads to significantly lower shear stress

Table 4.13 Shear strength and shear stress values at the interface [51]

Specimen	τ_{fud}: MPa	τ_x(for P_{max}): MPa	τ_x(for $P(s = 0.2\ \mathrm{mm})$): MPa
NSC-S-1	1.51	2.67	1.50
NSC-S-2	1.51	2.80	1.52
NSC-S-corr-1	1.51	1.65	1.14
NSC-S-corr-2	1.51	1.78	0.92
PVAFRGC-S-1	1.51	2.73	1.45
PVAFRGC-S-2	1.51	2.62	1.69
PVAFRGC-S-corr-1	1.51	2.76	2.00
PVAFRGC-S-corr-2	1.51	2.84	1.98

values for the corroded specimens strengthened with NSC (NSC-S-corr-1 and NSC-S-corr-2) compared with the respective values of the non-corroded specimens (NSC-S-1 and NSC-S-2).

Concluding remarks

The effectiveness of the use of additional PVAFRGC layers reinforced with steel bars for the improvement of the mechanical performance and durability of strengthened RC beams was examined in this section. Comparisons were made with corresponding specimens strengthened with conventional RC layers and the following conclusions were reached.

- The mass loss of the steel bars from corrosion was found to be much lower in specimens strengthened with PVAFRGC layers than in those strengthened with NSC layers, which demonstrates the improved resistance of PVAFRGC to steel corrosion.
- It is evident from the load–deflection results that corrosion had a significant negative impact on the maximum load for specimens strengthened with reinforced NSC layers, whereas the corrosion did not have any significant impact on the maximum load for specimens strengthened with PVAFRGC layers.
- Based on the interface slip values obtained from the bending tests, it was shown that specimens strengthened with reinforced PVAFRGC layers generally had lower maximum slip values than those obtained from specimens strengthened with reinforced NSC layers. This suggests that the specimens reinforced with PVAFRGC layers have enhanced interface conditions.
- Overall, it is clear from the experimental findings reported in this section that the application of a reinforced PVAFRGC layer for the strengthening of existing RC elements can result in increased corrosion resistance and noticeably improved structural performance.

4.5.3 Strengthening of URM walls using UHPFRC

The structural strengthening of URM structures is an emerging field worldwide. URM is one of the most popular construction types in the UK and in most countries. However, there are issues with the performance of these structures, in particular in earthquake-prone areas and in low- and middle-income countries. The main structural deficiencies of the URM structures, in particular under extreme loading conditions (for example, earthquakes and high winds), are linked to the poor mechanical qualities of the materials and, therefore, the majority of the existing URM structures are vulnerable to natural hazards. Additionally, many of these structures in low- and middle-income nations were built using substandard materials and improper methods, which places them at a very high risk.

This section focuses on the structural strengthening of URM walls using UHPFRC [57, 58].

The strengthening of existing URM using external high-strength materials is a challenging task owing to the poor connection conditions between the new materials and the existing substrate, which may cause debonding and early failure of the strengthened structures.

An area that is currently being researched is the use of innovative HPC for the structural improvement of the URM. Fibre-reinforced polymers (FRPs) have been explored for their external applications over the last few years [59–62] and it has been demonstrated that their superior tensile strength properties can help to improve the mechanical properties of URM. However, because of the weak connection between the URM and the FRPs, premature debonding and subsequent failures are frequently experienced.

In the last decade, there has been a lot of research carried out on the development of cementitious high-performance materials, and innovations in the development and applications of these materials have been proposed. For the flexural and shear strengthening of existing RC elements, the use of UHPFRC has been shown to be extremely effective [23, 46].

The effectiveness of the use of UHPFRC for the structural strengthening of URM walls for both in-plane and out-of-plane stress conditions has been numerically studied [63, 64] and the results show that the strength and the stiffness of the URM can be significantly enhanced by the addition of UHPFRC.

However, experimental studies in this field are very limited. In this section, the results of two experimental works on the strengthening of low-strength (aerated) brick walls [57] and on the strengthening of engineering brick walls [58] using UHPFRC are presented.

These studies use two different types of bricks, and extensive experimental investigation using UHPFRC layers with varying characteristics was conducted. Flexural out-of-plane tests were carried out on the URM walls after the application of UHPFRC layers of varying thicknesses with and without the addition of UHPFRC at the joints. The experimental findings show that the suggested technique can significantly increase both the ductility and ultimate load of the URM walls. The suggested method also provides better UHPFRC-to-URM interface conditions.

4.5.3.1 Description of the experimental investigation

The geometry and dimensions of the analysed specimens, along with the material parameters and a description of the UHPFRC preparation, are presented in this section for URM with both low-strength (aerated) bricks and engineering bricks.

URM with low-strength (aerated) bricks Low-strength aerated concrete bricks were used in this investigation to simulate the behaviour of typical URM structures in low- and middle-income areas. The bricks are 65 mm by 100 mm by 215 mm in size, and have compressive strength equal to 3.57 MPa. Eight specimens were examined in total, including two URM specimens (URM-1 and URM-2), two URM specimens strengthened with a 14 mm thick UHPFRC layer (UHPFRC_14 mm-1, UHPFRC_14 mm-2), two URM specimens strengthened with a 22 mm thick UHPFRC layer (UHPFRC_22 mm-1, UHPFRC_22 mm-2) and two URM specimens strengthened with a 30 mm thick UHPFRC layer (UHPFRC_30 mm-1, UHPFRC_30 mm-2) [57].

Figure 4.64 shows the size of the URM specimens as well as one of the experimental samples before testing.

Cement, lime and sand were used for the mortar at the joints of the examined specimens in a volume ratio of 1:1/3:3, with a water to cement weight ratio of 0.32, which is typical for mortar used in URM walls. After carrying out the required compressive cube tests, it was found that this mortar had a compressive strength of 21.24 MPa.

Figure 4.64 Geometry and experimental URM specimen with low-strength (aerated) bricks before testing [57]

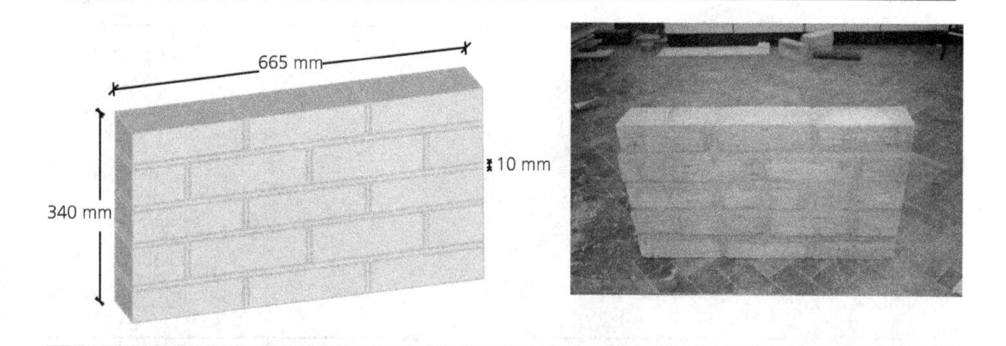

UHPFRC layers with three different thicknesses (14 mm, 22 mm and 30 mm) were cast in connection to the existing specimens and the strengthened URM specimens are presented in Figure 4.65.

Table 4.14 presents the UHPFRC mix composition which was used in this study.

A mixture of silica sand with a maximum particle size of 0.5 mm, dry silica fume, GGBS and cement (32.5R) was used. After mixing of the dry components, water and a polycarboxylate superplasticiser were added to the mixture, followed by the addition of straight steel fibres with a length of 6 mm, diameter of 0.16 mm and a tensile strength of 3000 MPa.

After the casting of the UHPFRC layers, the specimens were allowed to cure at ambient conditions for two months before their out-of-plane testing. The experimental set-up of the

Figure 4.65 Strengthened URM specimens with additional UHPFRC layers [57]

Table 4.14 UHPFRC mix composition [57]

Material	Mix proportions: kg/m^3
Cement	657
GGBS	418
Silica fume	119
Silica sand	1051
Superplasticisers	59
Water	185
Steel fibres (3%)	235.5

out-of-plane tests is presented in Figure 4.66. Tests were carried out under displacement control with a loading rate of 0.1 mm/min using a span length of 520 mm. The digital image correlation (DIC) system was used to monitor the strains and cracks at the interface between the UHPFRC and the URM in all of the strengthened units, and the load against displacement was also recorded during the testing (Figure 4.66).

To determine the compressive strength and tensile stress–strain properties of the UHPFRC, dog-bone specimens with a cross-section of 14 mm by 20 mm and cubes with 100 mm sides were also prepared and tested on the same day as the testing of the URM specimens. The maximum tensile strength of the cube was determined to be almost equal to 4 MPa, whereas the compressive strength of the cube was found to be 116 MPa.

URM with engineering bricks The use of class B engineering bricks for the construction of the initial URM brick walls is examined in this section. These bricks had dimensions of 65 mm by 102.5 mm by 215 mm and a compressive strength of 50 MPa. Sand and cement were used for the mortar of these brick walls with a volume ratio of one to four, and the water to cement

Figure 4.66 (a) Experimental set-up of the out-of-plane tests and (b) the DIC monitoring system for the URM-to-UHPFRC interface [57]

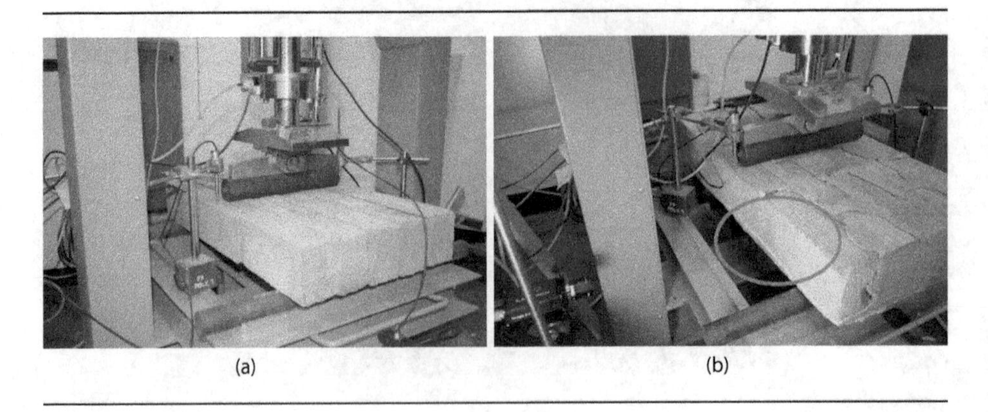

(a) (b)

weight ratio was equal to 0.5. The thickness of the joints was 10 mm to represent a typical URM wall construction (Figure 4.67).

The mean compressive strength of the mortar, which was measured using cubes with 100 mm sides, was determined to be 9.36 MPa. In this study, 17 URM brick walls in total were studied. Table 4.15 gives a description of some of the key characteristics of the examined specimens.

Three URM specimens were tested without strengthening (URM), three specimens were strengthened with a 10 mm layer and 3% of steel fibres (UHPFRC-10-3%), two specimens were strengthened with a 30 mm layer and 3% of steel fibres (UHPFRC-30-3%), three specimens were strengthened with a 30 mm layer and 1% of steel fibres (UHPFRC-30-1%), and three specimens were strengthened with a 30 mm layer and 6% of steel fibres (UHPFRC-30-6%). For the remaining three specimens, a new application was tested in which the old conventional mortar was removed at a depth equivalent to one-third of the wall's overall thickness, and UHPFRC was used for repointing in addition to a layer of 30 mm UHPFRC with 3% fibres (UHPFRC-30-REP-3%).

The URM specimens were strengthened 28 days after their casting, and they underwent a second two-month curing period at a constant temperature of 21–23°C and relative humidity of 55–60%. The UHPFRC that was used for the strengthening of these specimens was the same

Figure 4.67 Geometry and experimental URM specimen with engineering bricks before testing [58]

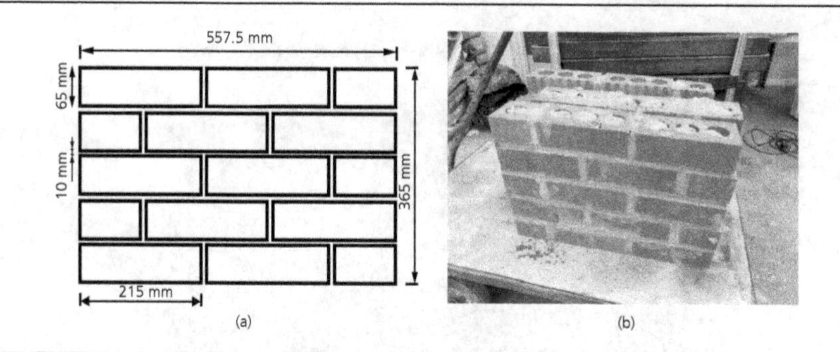

Table 4.15 Key characteristics of the examined specimens [58]

Specimens ID	Number of specimens	Thickness of UHPFRC layer: mm	UHPFRC fibre volume fraction: %
URM	3	N/A	N/A
UHPFRC-10-3%	3	10	3
UHPFRC-30-3%	2	30	3
UHPFRC-30-REP-3%	3	30 and repointing	3
UHPFRC-30-1%	3	30	1
UHPFRC-30-6%	3	30	6

as that used for strengthening of the URM with the low-strength (aerated) bricks (Table 4.14). In this study, two distinct types of fibres were used. For specimens UHPFRC-10-3% and UHPFRC-30-REP-3%, fibres with a tensile strength of 2750 MPa, nominal diameter 0.21 mm and length 13 mm were used. For all of the other strengthened specimens of Table 4.15, fibres with nominal diameter 0.15 mm, length 6 mm and tensile strength 3000 MPa were used.

Specimens were placed in the moulds for the casting of the UHPFRC layers, and UHPFRC was then poured at the required levels according to the layer thickness (Figure 4.68).

Figure 4.68 URM walls strengthened with UHPFRC [58]

(a)

(b)

(c)

One of the parameters that needs to be examined in future studies is the flowability of UHPFRC, which needs to be adjusted to allow casting of the layers with rendering of the walls in a vertical configuration since this procedure is not representative of the actual construction technique used in practice.

The compressive and tensile capabilities of UHPFRC were assessed using cubes with 100 mm sides, and dog-bone-shaped specimens having a 20 mm by 14 mm cross-section (Figure 4.69).

To assess the out-of-plane performance of the URM walls, a three-point loading testing system with a span length of 500 mm was chosen (Figures 4.70 and 4.71), which is similar to that presented in Figure 4.66 for testing the low-strength URM walls.

The mid-span deflection was measured using two LVDTs in the two sides of the specimens, and displacement control was employed to test the walls at a rate of 0.1 mm/min. In addition to the monitoring of the strains and cracks at the interface between the UHPFRC and the URM using a DIC system (Figure 4.71) the load versus mid-span deflection was also measured using two LVDTs (Figure 4.71).

Figure 4.69 Presentation of the examined direct tensile (dog-bone) specimens [58]

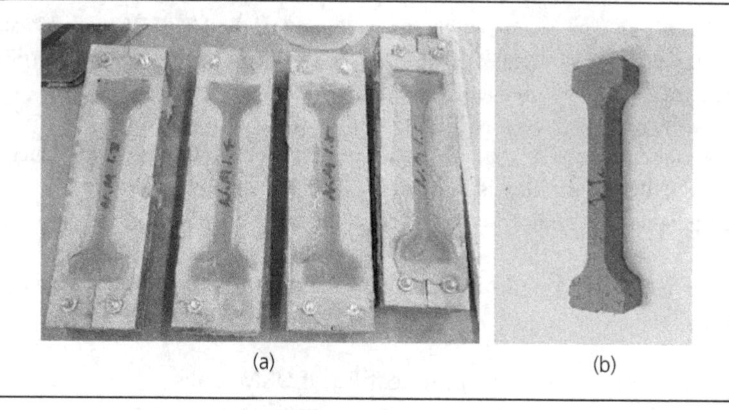

(a)

(b)

Figure 4.70 Representation of the testing set-up [58]

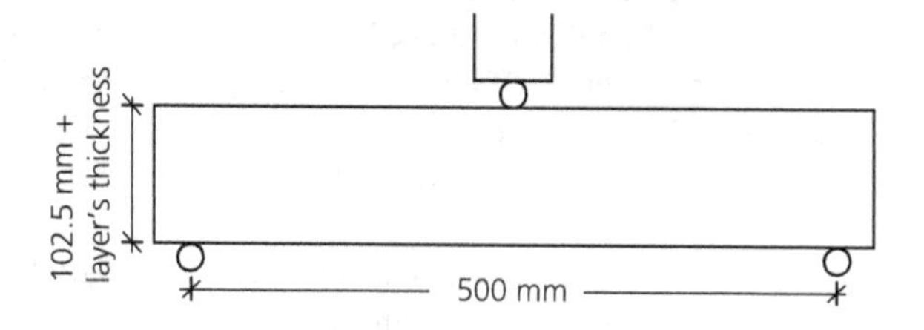

Figure 4.71 Experimental set-up and monitoring of the interface using a DIC system [58]

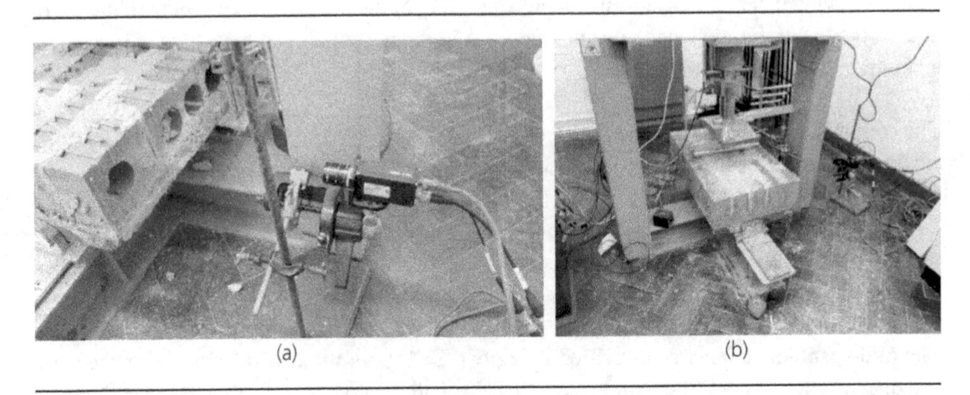

(a) (b)

To determine the UHPFRC compressive strength and tensile stress–strain properties, dog-bone specimens with a cross-section of 14 mm by 20 mm and cubes with 100 mm sides were tested on the same day as the testing of the strengthened URM walls. In many of the examined dog-bone specimens, failure occurred near the grips and these specimens were eliminated from the calculation of the UHPFRC characteristics. The maximum tensile strength of the examined specimens that were successfully tested was found to be around 4 MPa.

With regard to the compressive strength, it was found that the UHPFRC with 1% steel fibres had a mean compressive strength of 104 MPa, whereas the respective values with 3% and 6% fibres were found to be 133 MPa and 124 MPa, respectively.

Both tensile and compressive strength characteristics were lower than the values typically expected for UHPFRC and this is attributed to the low-strength concrete (32.5R) and the ambient temperature of curing that were used in this investigation.

The experimental results for all of the examined specimens are presented in the following section (4.5.3.2).

4.5.3.2 Results of the out-of-plane testing of URM walls

In this section, the experimental results together with some analytical calculations for the examined URM walls with low strength (aerated) and with engineering bricks are presented.

Results of URM walls with low-strength (aerated) bricks The load against mid-span displacement results of all the examined URM walls with low-strength (aerated) bricks are presented in Figure 4.72. Also, it should be mentioned that an error was introduced in the initial displacement recordings of one of the UHPFRC_14 mm and one of the UHPFRC_22 mm specimens, due to undesirable displacements at the supports, and these recordings were corrected using the respective values of the other identical specimens of each pair.

Based on the results of Figure 4.72, it is evident that the addition of UHPFRC layers significantly increases maximum load, and that maximum load capacity increases with layer

Figure 4.72 Experimental load–deflection results for the out-of-plane tests of all the examined URM walls with low-strength (aerated) bricks [57]

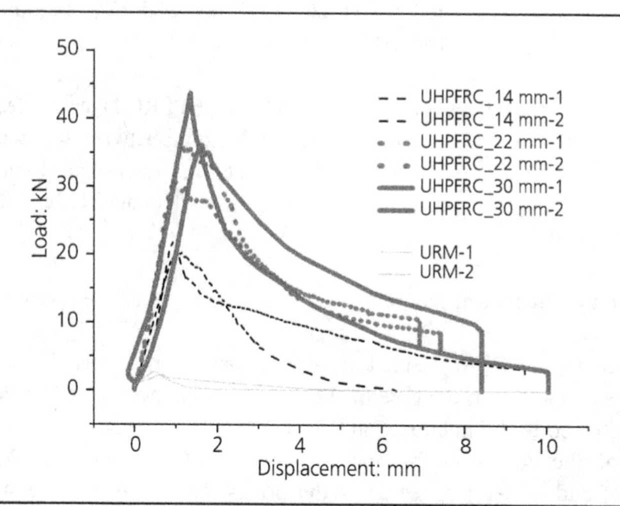

thickness increment, as expected. In addition to the ultimate load capacity enhancement, the area under the load–displacement graphs and, consequently, the energy absorption of the strengthened specimens are significantly improved. This is attributed to the superior tensile stress–strain characteristics of UHPFRC and to the bridging action of the fibres. The average load-displacement findings of the two identical specimens of each of the examined types are illustrated in Figure 4.73.

Figure 4.73 Average load–deflection results for the out-of-plane tests of all the examined URM walls with low-strength (aerated) bricks [57]

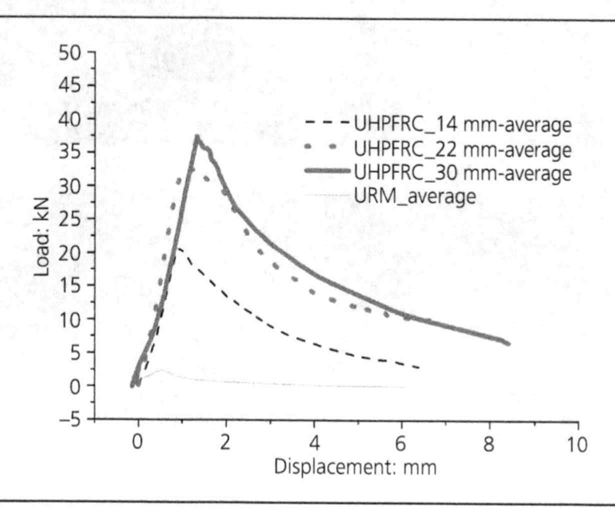

According to the findings of Figure 4.73, the maximum average load capacity of URM was determined to be 2.37 kN. The maximum load values of the strengthened specimens were much higher and were calculated to be 20.86 kN for UHPFRC_14 mm, 32.56 kN for UHPFRC_22 mm and 37.44 kN for UHPFRC_30 mm.

With regard to the interface between the bricks and the UHPFRC layer, the tests show a very good connection between the two materials. Perfect bonding between the two materials was achieved up to the point where cracks started to appear on the outermost tensile side of the bricks, which were then propagated to the UHPFRC-to-bricks interface, and finally appeared on the UHPFRC layer.

Figure 4.74 displays the typical failure modes of all four types of the examined specimens.

URM specimens (Figure 4.74(a)) failed in a brittle manner as expected, and at a relatively low ultimate load capacity. In the case of the strengthened specimens, the failure started at the bricks, progressed to the interface, and subsequently propagated to the UHPFRC layer. The majority of the cracks in the specimens UHPFRC_14 mm and UHPFRC_22 mm (Figures 4.74(b) and 4.74(c)) appeared at the bricks and the interface rather than at the UHPFRC layer, whereas in the UHPFRC_30 mm specimen (Figure 4.74(d)), a significant crack opening was observed at the middle of the span of the UHPFRC in addition to significant brick damage.

Figure 4.74 Illustration of typical modes of failure of (a) URM, (b) UHPFRC_14 mm, (c) UHPFRC_22 mm and (d) UHPFRC_30 mm [57]

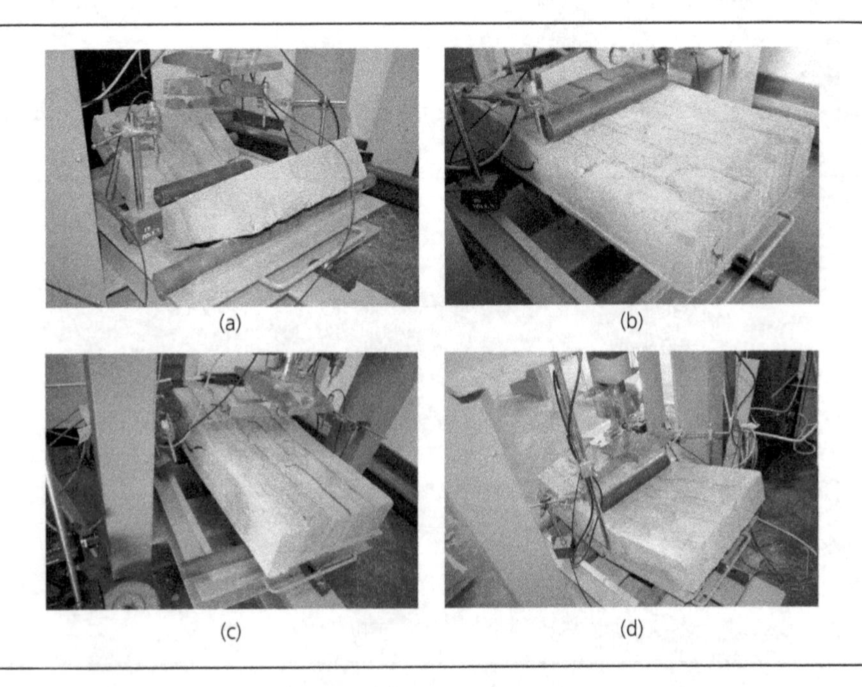

(a) (b) (c) (d)

It is crucial to note that, in all of the examined strengthened specimens, almost perfect bond conditions were observed at the UHPFRC-to-bricks interface up to the maximum load, and any interface slips and cracks appeared at the post-peak region and in the softening branch of the load–deflection curve.

The DIC system, which was employed in this work to monitor the interface, was used to conduct a thorough assessment of the interface behaviour. The shear strain was calculated for one of each type of strengthened specimen (UHPFRC_14 mm, UHPFRC_22 mm and UHPFRC_30 mm).

Figure 4.75 shows indicative results for the shear strain distribution at the middle and at the end of the imposed mid-span displacement for UHPFRC_14 mm.

According to the results of Figure 4.75, there are high shear strain values and slip at the interface for the specimen strengthened with a 14 mm thick UHPFRC layer (UHPFRC_14 mm), not only at the end of the loading history (Figure 4.75(b)), but also for the loading stage in the middle of the load–displacement history (Figure 4.75(a)).

Figures 4.76 and 4.77 show the corresponding shear strain distributions for the UHPFRC_22 mm and UHPFRC_30 mm specimens. Based on the results of these two types of strengthened specimens, when 22 mm and 30 mm thick UHPFRC layers were applied, there were no significant shear strain values during the loading stage at the middle of the maximum imposed displacement. This shows a very strong interface connection even at a relatively late loading stage, which corresponds to the softening branch of the load–displacement graph where the bricks have sustained severe damage.

Figure 4.75 Distribution of shear strains at the interface of specimen UHPFRC_14 mm for two loading stages: (a) the middle and (b) the maximum imposed displacement [57]

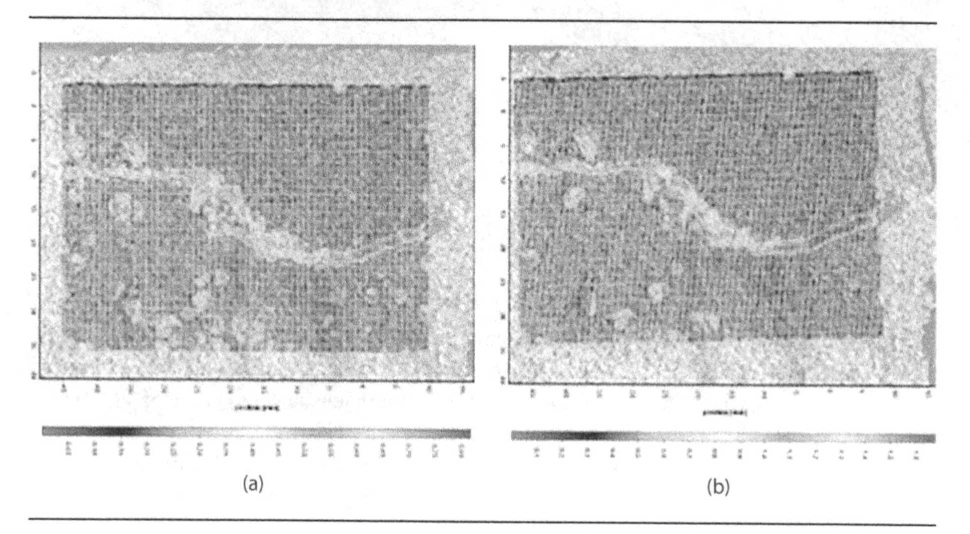

Figure 4.76 Distribution of shear strains at the interface of specimen UHPFRC_22 mm for two loading stages: (a) the middle and (b) the maximum imposed displacement [57]

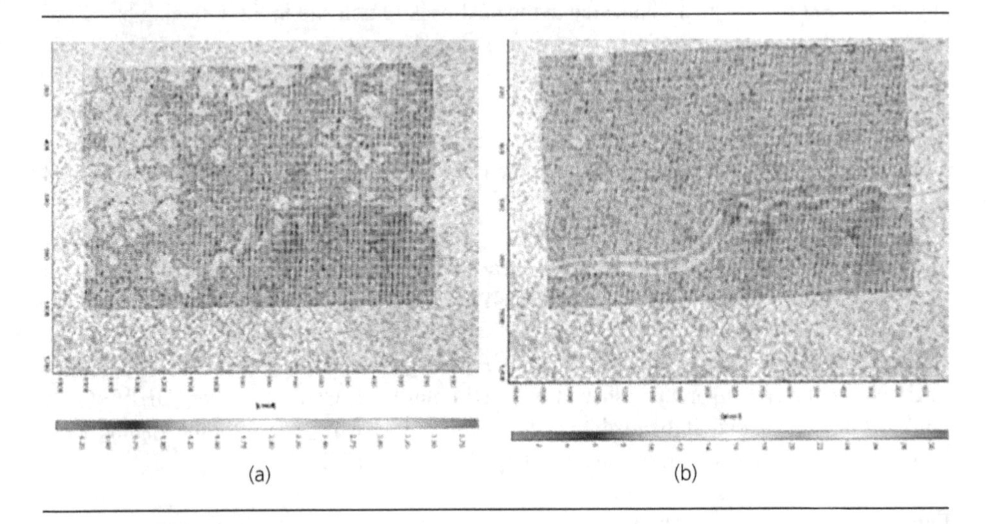

Figure 4.77 Distribution of shear strains at the interface of specimen UHPFRC_30 mm for two loading stages: (a) the middle and (b) the maximum imposed displacement [57]

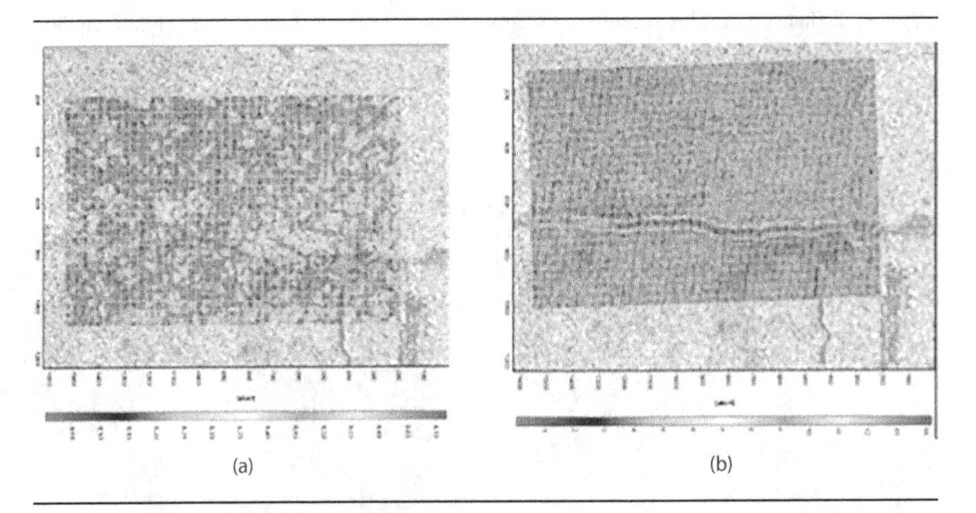

To quantify the contribution of the additional UHPFRC layer to the maximum load capacity of the strengthened specimens, a simplified analytical approach is presented. Cross-sectional analysis was performed using the direct (dog-bone) tensile testing results, which were obtained experimentally. Based on these results, a uniform maximum tensile stress value of 4 MPa was used for the UHPFRC layer. The internal force F_{S_UHPFRC} (Figure 4.78) was subsequently determined by multiplying the maximum stress value by the thickness and width of the layer. The internal force F_{S_UHPFRC} was then multiplied by the depth, which was taken equal to the

Figure 4.78 Internal force distribution used for the cross-sectional analysis of the strengthened URM specimens [57]

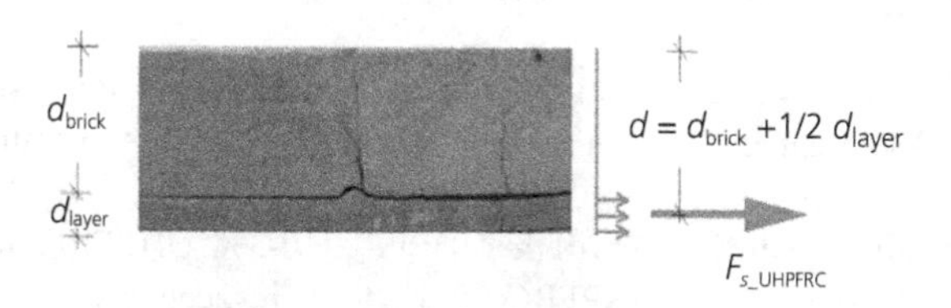

total depth of the brick and half of the depth of the layer, to calculate the moment increment (Figure 4.78).

The ultimate load was analytically calculated using the cross-sectional analysis (Figure 4.78) for all three types of strengthened URM walls. The analytical ultimate load values for UHPFRC_14 mm, UHPFRC_22 mm and UHPFRC_30 mm were calculated to be 18.04 kN, 27.92 kN and 38.46 kN respectively. Figure 4.79 shows comparisons between the calculated values and the values acquired experimentally.

The results of Figure 4.79 show that there is a very good agreement between the maximum load values predicted using the simplified analytical method and the corresponding experimental findings. The UHPFRC and URM interface showed excellent bond conditions for the loading stage up to the ultimate load capacity, as previously stated. As a result, the simplified analytical procedure presented here, which assumes perfect connection between the two materials, is appropriate for the estimation of the contribution of the additional UHPFRC layer to the ultimate load capacity.

Figure 4.79 Analytical against experimental results for the ultimate load capacity of the examined URM with low-strength (aerated) bricks [57]

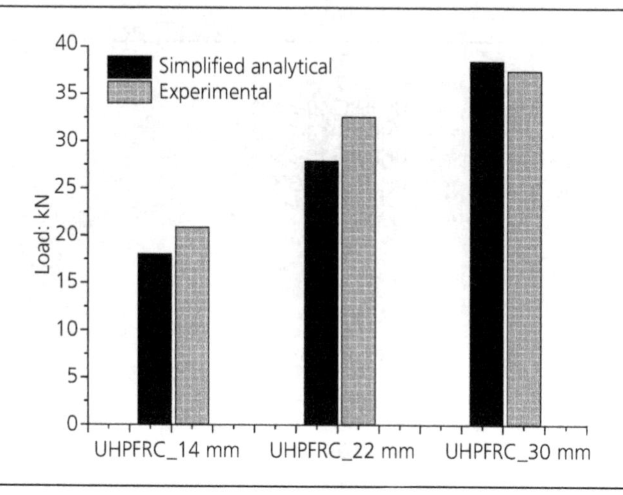

Results of URM walls with engineering bricks Figures 4.80–4.85 show the average load against mid-span displacement results together with the individual results for all of the examined specimens with engineering bricks.

Figure 4.80 displays the average results as well as the load–deflection results for the three URM specimens. As anticipated, brittle failure occurred in all of the examined specimens.

Figures 4.81–4.85 illustrate the findings of the UHPFRC-10-3%, UHPFRC-30-1%, UHPFRC-30-3%, UHPFRC-30-Rep-3% and UHPFRC-30-6% tests, respectively.

Figure 4.80 Experimental load–deflection results for the out-of-plane tests of the examined URM specimens [58]

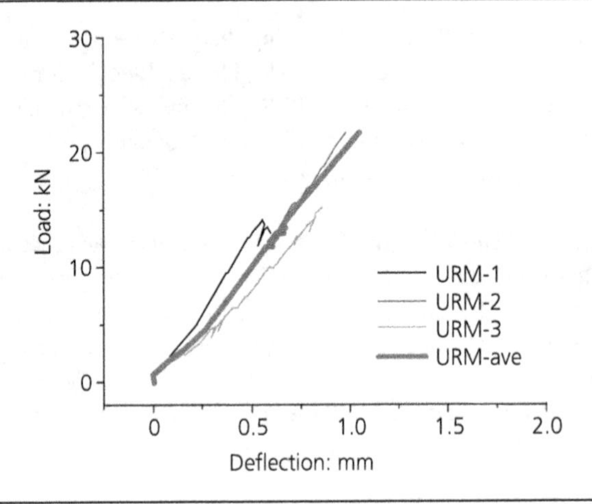

Figure 4.81 Experimental load–deflection results for the out-of-plane tests of the examined UHPFRC-10-3% specimens [58]

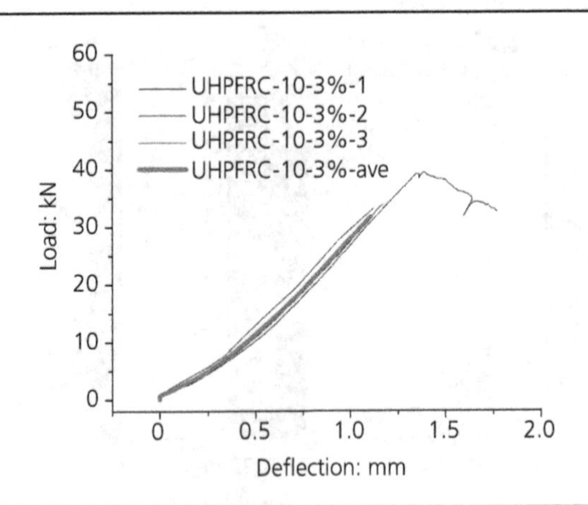

Figure 4.82 Experimental load–deflection results for the out-of-plane tests of the examined UHPFRC-30-1% specimens [58]

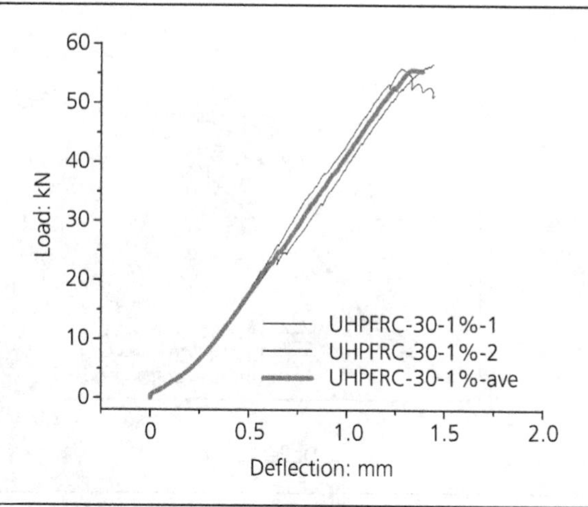

Figure 4.83 Experimental load–deflection results for the out-of-plane tests of the examined UHPFRC-30-3% specimens [58]

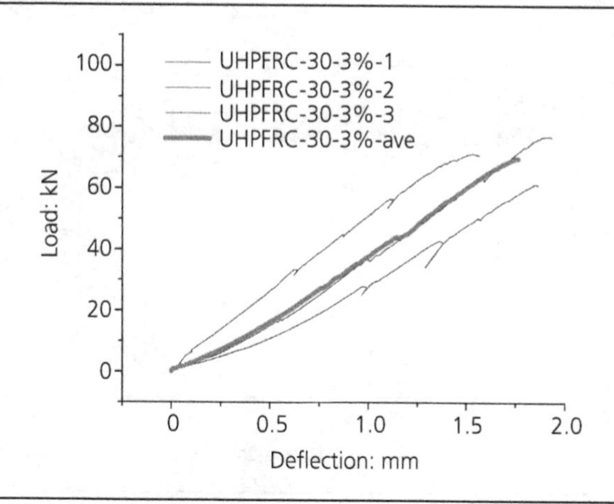

Figure 4.86 compares the average load against mid-span displacement results for all of the examined specimens with engineering bricks.

It is evident from the results of Figure 4.86 that the addition of UHPFRC significantly increased the strength and stiffness of the URM specimens.

Figure 4.84 Experimental load–deflection results for the out-of-plane tests of the examined UHPFRC-30-Rep-3% specimens [58]

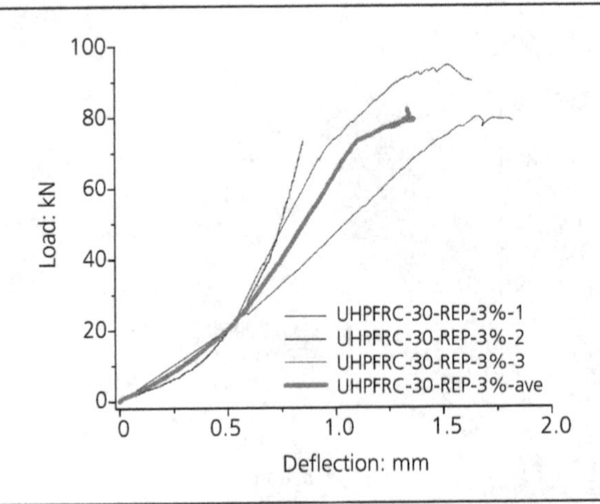

Figure 4.85 Experimental load–deflection results for the out-of-plane tests of the examined UHPFRC-30-6% specimens [58]

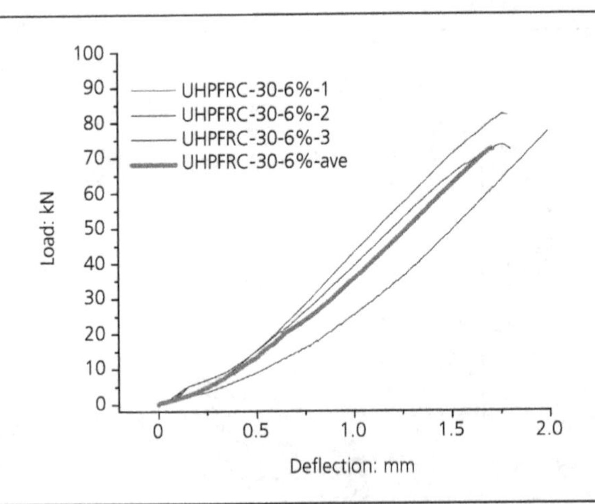

All of the URM specimens showed brittle failure, with cracks beginning at mortar joints and extending, as expected, to the bricks.

With regard to the strengthened specimens, cracks started in the brick walls, spread to the brick-to-UHPFRC interface, and finally reached the layers of UHPFRC. All of the examined types of strengthened elements experienced a similar failure mechanism.

Figures 4.87(a)–4.87(e) show typical failure modes for each type of the examined specimens.

Figure 4.86 Experimental load–deflection results for the out-of-plane tests of the average results of all the examined types [58]

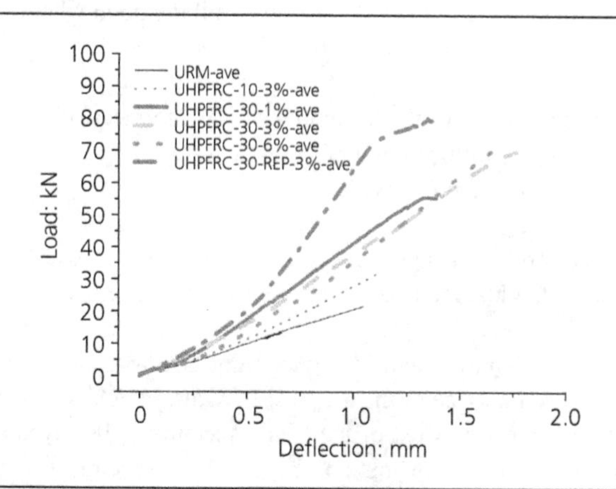

Figure 4.87 Characteristic failure modes for specimens (a) UHPFRC-30-1, (b) UHPFRC-10-3, (c) UHPFRC-30-3, (d) UHPFRC-30-6 and (e) UHPFRC-30-3-REP [58]

The experimental results showed that there is a very strong bond at the interface between the bricks and the UHPFRC. An almost perfect interface connection was observed for all of the examined specimens up until a point near failure and until the point where cracks spread from the bricks to the interface and then to the UHPFRC layer.

A DIC system was used to monitor the behaviour of the interfaces of all of the strengthened specimens, and the results confirmed the perfect bond at the URM-to-UHPFRC interfaces up until a point close to the maximum stress.

Figure 4.88 shows results of the shear strain distribution for one of the tested UHPFRC-30-3% specimens at the end of the loading (maximum imposed displacement).

Figure 4.88 clearly shows that even when the maximum load was surpassed, there is no shear strain concentration at the interface of the tested specimens, which is an indication of perfect bond conditions at the interface. Most of the tested specimens displayed similar behaviour. Interface slips and cracks occurred in some of the examined specimens at a very late loading stage and at the post-peak behaviour of the load–deflection graphs, when cracks were propagated to the level of the interface.

Figure 4.89 shows indicative DIC data for one of the UHPFRC-30-6% specimens before and after the concentration of high interface shear strains.

These two images display the shear strain distributions at a loading stage near the end (Figure 4.89(a)) and at the end of the test (Figure 4.89(b)) and, more specifically, before (Figure 4.89(a)) and after (Figure 4.89(b)) the crack propagation from the bricks to the UHPFRC-to-bricks interface.

Figure 4.88 Distribution of shear strains at the interface of specimen UHPFRC-30-3% at the maximum imposed displacement [58]

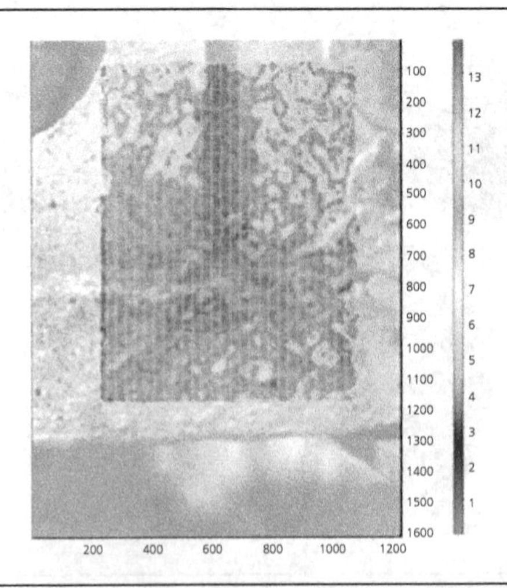

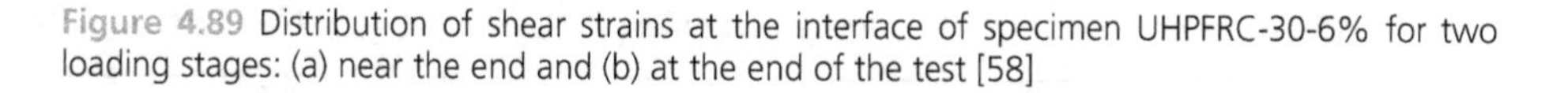

Figure 4.89 Distribution of shear strains at the interface of specimen UHPFRC-30-6% for two loading stages: (a) near the end and (b) at the end of the test [58]

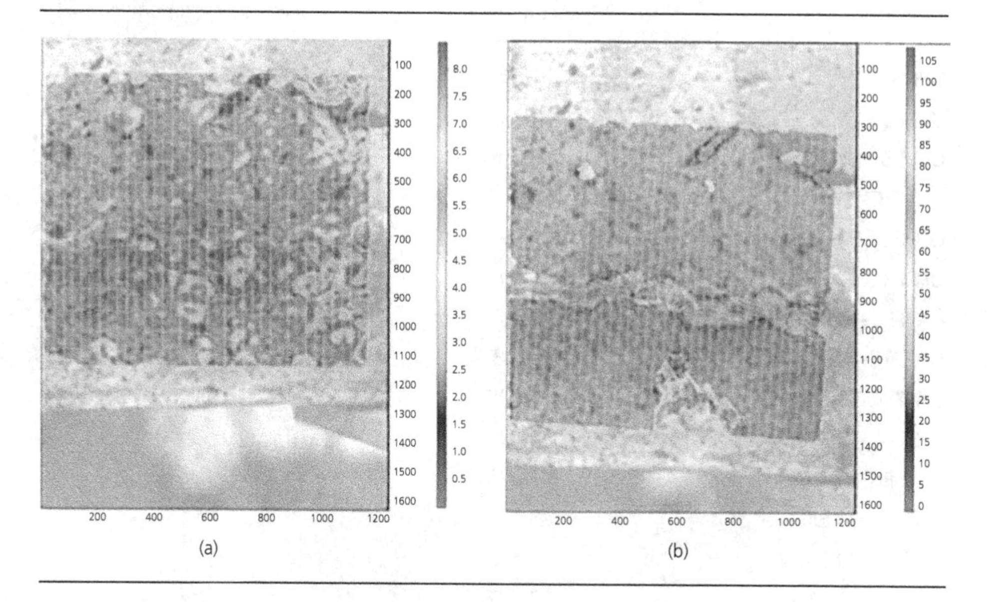

Table 4.16 displays the ultimate load results for all of the examined specimens, which were obtained experimentally from the average load–displacement results.

The results of Table 4.16 show that the maximum load was increased by 110% for the specimen strengthened with a 10 mm layer with 3% fibre percentage (UHPFRC-10-3%), whereas an increment of 232% was achieved when a layer of thickness 30 mm (UHPFRC-30-3%) was applied. The use of UHPFRC with 6% steel fibres and a 30 mm layer (UHPFRC-30-6%) led to an even higher load increment, which was found to be 379%. Partial repointing, which involved filling 1/3 of the joint depth with UHPFRC with 3% steel fibres, led to the highest load increment of 388%.

Following the simplified cross-sectional analysis (Figure 4.78) analytical calculations were conducted assuming perfect bonding at the interface. Considering that the tensile stress of UHPFRC is equivalent to 4 MPa, analytical calculations of the additional load were carried out for UHPFRC-10%, UHPFRC-30% and UHPFRC-30-REP-3%.

For the repointed specimens (UHPFRC-30-REP-3%), the addition of UHPFRC led to the further improvement of the tensile side of the walls. For the cross-sectional analysis of this sample, it was assumed that the UHPFRC stress block extends until the point where UHPFRC was used at the joints (about one-third of the depth).

Figure 4.90 presents a comparison between the theoretical and the experimental results.

Table 4.16 Ultimate load results of the examined specimens [58]

Specimen	Maximum load: kN	Average maximum load: kN
URM-1	15.1	16.93
URM-2	14.02	
URM-3	21.66	
UHPFRC-30-1%-1	56.623	56.26
UHPFRC-30-1%-2	55.9	
UHPFRC-10-3%-1	33.28	35.55
UHPFRC-10-3%-2	33.9	
UHPFRC-10-3%-3	39.48	
UHPFRC-30-3%-1	77.105	70.02
UHPFRC-30-3%-2	61.529	
UHPFRC-30-3%-3	71.423	
UHPFRC-30-6%-1	73.421	81.11
UHPFRC-30-6%-2	87.672	
UHPFRC-30-3%-3	82.237	
UHPFRC-30-Rep-3%-1	73.34	82.61
UHPFRC-30-Rep-3%-2	94.49	
UHPFRC-30-Rep-3%-3	80.01	

The results of Figure 4.90 show that there is a good degree of agreement between the experimental and the analytical findings. The assumption that the UHPFRC stress block of the repointed samples (UHPFRC-30-3-REP) should be extended to the depth where repointing was performed appears to lead to results which are in good agreement with the respective experimental ones.

4.5.3.3 Concluding remarks

In this section, the results of an extensive experimental programme are presented on URM strengthened with UHPFRC. The strengthening of two different types of URM walls (with low-strength bricks and with engineering bricks) was studied to assess the effectiveness of the use of additional UHPFRC with varying thicknesses and different amount of steel fibres.

The load–deflection results were recorded and the interface was monitored using a DIC system. In addition to the experimental results, a simplified analytical approach was presented for calculating the ultimate load enhancement of the strengthened specimen.

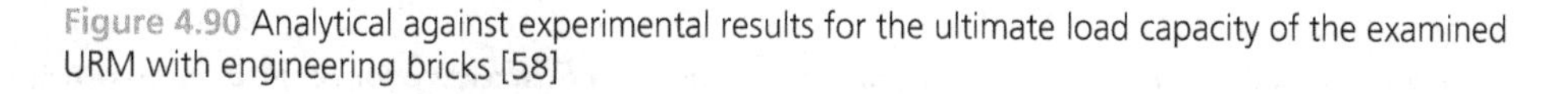

Figure 4.90 Analytical against experimental results for the ultimate load capacity of the examined URM with engineering bricks [58]

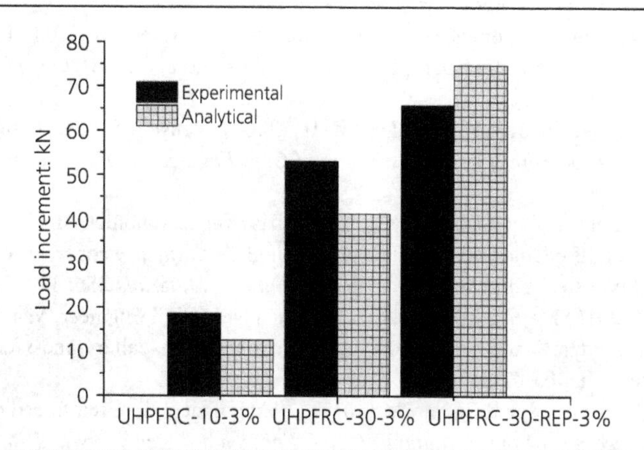

The following main conclusions were drawn.

- The addition of UHPFRC layers led to significantly enhanced ultimate load capacity and energy absorption in all of the examined URM specimens (with low-strength bricks and with engineering bricks).
- An increment in the thickness of the additional layer led to further enhancements of both strength and ductility.
- In the case of strengthened specimens, the failure started with the initiation of cracks at the bricks, which then progressed to the interface, and finally cracks appeared at the UHPFRC layers.
- Perfect bonding at the URM-to-UHPFRC interface was observed (using the DIC system) until the ultimate load value in all of the examined specimens (with low-strength bricks and with engineering bricks). After the maximum load and when cracks propagated from the bricks to the interface, high shear strains appeared at the interface, which led to the development of interface slips and cracks.
- The proposed simplified analytical method using cross-sectional analysis and assuming perfect bonding at the URM-to-UHPFRC interface shows very good agreement with the corresponding experimental results.
- In the case of URM with engineering bricks, partial repointing of the existing mortar was also examined in addition to the UHPFRC layer. This technique led to further enhancement of the structural performance of the strengthened URM walls; therefore, this technique should be considered, in particular, in cases of defective mortar at the joints.

REFERENCES

1. Ecolanes (2009) *Economical and Sustainable Pavement Infrastructure for Surface Transport.* Final Activity Report for EcoLanes FP6 STREP Project.
2. Neocleous K, Angelakopoulos H, Pilakoutas K and Guadagnini M (2011) Fibre-reinforced roller-compacted concrete transport pavements. *Proceedings of the Institute of Civil Engineers, Transport* **164**: 97–109.
3. Pilakoutas K, Neocleous K and Tlemat H (2004) Reuse of steel fibres as concrete reinforcement. *Proceedings of the Institution of Civil Engineers – Engineering Sustainability* **157(3)**: 131–138.
4. Isa MN, Pilakoutas K, Guadagnini M and Harris Angelakopoulos H (2020) Mechanical performance of affordable and eco-efficient ultra-high performance concrete (UHPC) containing recycled tyre steel fibres. *Construction and Building Materials* **255**: 119272.
5. Hansford M (2015) Major Milestone for Lee Tunnel New Civil Engineer. *New Civil Engineer.* 3 Feb. 2015. https://www.newcivilengineer.com/archive/the-gallery-and-video-major-milestone-for-lee-tunnel-03-02-2015/
6. Johnston RP, Psomas S and Eddie CM (2017) Design of steel fibre reinforced concrete tunnel linings. *Proceedings of the Institution of Civil Engineers – Structures and Buildings* **170(2)**: 115–130.
7. The Concrete Society (TCS) (2007) *Guidance for the Design of Steel-fibre-reinforced Concrete.* The Concrete Society, Camberley, UK, Technical Report TR63.
8. Mitchell R (2011) The Lee Tunnel, *Wastewater Treatment and Sewerage*, pp. 57–62.
9. Costes E, Jewel P, Michel C and Pogu F (208) Lee Tunnel project – the first step toward a cleaner River Thames. *Proceedings of the Institution of Civil Engineers – Civil Engineering* **171(2)**: 69–76.
10. Liwanag V (2021) The UHPFRC footbridge in Lužec nad Vltavou. Emagazine by ArchiExpo.
11. Archiscience (2021) Footbridge in Příbor by Petr Tej, Marek Blank & Jan Mourek: As a Stone Laid Over the Water. https://www.archdaily.com/961263/footbridge-in-pribor-petr-tej-plus-marek-blank-plus-jan-mourek (accessed 26/09/2023).
12. Blank M, Tej P, Kolísko J and Vráblík L (2016) Design of experimental suspended footbridge with deck made of UHPC. *MATEC Web of Conferences* **77**: 08005.
13. Newtson CM, Weldon BD, Toledo WK, Alvarez A and Manning MP (2021) *Field Implementation and Monitoring of an Ultra-High Performance Concrete Bridge Deck Overlay, Transportation Consortium of South-Central States*, Project No. 19CNMS01, Lead University: New Mexico State University, USA.
14. Goonewardena J, Ghabraie K and Subhani M (2020) Flexural performance of FRP-reinforced geopolymer concrete beam. *Journal of Composite Science* **4(4)**: 187.
15. Bligh R and Glasby T (2014) Geopolymer Precast Floor Panels: Sustainable Concrete for Australia's Global Change Institute. *Structure Magazine*. Jan 2014. https://www.structuremag.org/?p=1347 (accessed 26/09/2023).
16. Aldred JA (2013) Engineering properties of a proprietary premixed geopolymer concrete. *Proceedings of the Concrete Institute of Australia Biennial Conference, Concrete Conference*, paper no. 75.
17. Bligh R and Glasby T (2013) Development of geopolymer floor panels for the Global Change Institute at the University of Queensland. *Concrete in Australia* **40(1)**: 39–43.
18. Lampropoulos A, Tsioulou O, Paschalis S and Dritsos S (2021) Strengthening of reinforced concrete beams: RC versus UHPFRC layers. *IABSE Congress, Christchurch, New Zealand.* International Association for Bridge and Structural Engineering (IABSE).

19. Tsioulou O, Lampropoulos A and Dritsos S (2012) Experimental investigation of interface behaviour of RC beams strengthened with concrete layers. *Construction and Building Materials* **40**: 50–59.
20. Paschalis S, Lampropoulos A and Tsioulou O (2018) Experimental and numerical study of the performance of ultra high performance fiber reinforced concrete for the flexural strengthening of full scale reinforced concrete members. *Construction and Building Materials* **186**: 351–366.
21. Lampropoulos A, Tsioulou O, Nicolaides D and Petrou MF (2023) Strengthening of existing structures with UHPFRC: concrete-to-UHPFRC interfaces. Building for the future: durable, sustainable, resilient. *Proceedings of the fib Symposium 2023, Istanbul, Turkey.* (Ilki A, Çavunt D and Çavunt YS (eds)) Springer.
22. Farhat FA, Nicolaides D, Kanellopoulos A and Karihaloo BL (2006) High performance fibre-reinforced cementitious composite (CARDIFRC) – performance and application to retrofitting. *Engineering Fracture Mechanics* **74**: 151–167.
23. Lampropoulos A, Paschalis S, Tsioulou O and Dritsos S (2015) Strengthening of reinforced concrete beams using ultra high performance fibers reinforced concrete (UHPFRC). *Engineering Structures* **106**: 370–384.
24. Bruhwiler E and Denarie E (2008) Rehabilitation of concrete structures using ultra-high performance fibre reinforced concrete. In: *The Second International Symposium on Ultra High Performance*, Kassel, Germany (Fehling E, Schmidt M and Sturwald S (eds)) Kassel University Press GmbH, Kassel, Germany.
25. Al-Osta M, Isa M, Baluch M and Rahman M (2017) Flexural behaviour of reinforced concrete beams strengthened with ultra-high performance fiber reinforced concrete. *Construction and Building Materials* **134**: 279–296.
26. Safdar M, Takashi MT and Kakuma K (2016) Flexural behavior of reinforced concrete beams repaired with ultra high performance fiber reinforced concrete (UHPFRC). *Composite Structures* **157**: 448–460.
27. Paschalis SA and Lampropoulos AP (2021) Developments in the use of ultra high performance fiber reinforced concrete as strengthening material. *Engineering Structures* **233**: 111914.
28. Ridström HLA (2019) *Investigation of the Behavior of Ultra-High Performance Fibre Reinforced Concrete (UHPFRC) to Normal Concrete Interfaces*. Final year project (BEng), University of Brighton, UK.
29. Greek Code of Interventions (KAN.EPE.) (2013), Earthquake Planning and Protection Organization, Greece.
30. Lampropoulos A, Nicolaides D, Paschalis S and Tsioulou O (2021) Experimental and numerical investigation on the size effect of ultra-high performance fibre reinforced concrete (UHPFRC). *Materials, MDPI* **14(19)**: 5714.
31. Nicolaides D (2004) *Fracture and Fatigue of CARDIFRC*. PhD thesis, Cardiff University, Wales, UK.
32. Nicolaides D, Kanellopoulos A and Karihaloo BL (2006) Investigation of the effect of fibre distribution on the fatigue performance and the autogenous shrinkage of CAR-DIFRC. In: *Measuring, Monitoring and Modelling Concrete Properties* (Konsta-Gdoutos MS (ed.)) Springer Netherlands, Dordrecht, the Netherlands, pp. 3–16.
33. Nicolaides D, Kanellopoulos A, Petrou M, Savva P and Mina A (2015) Development of a new ultra high performance fibre reinforced cementitious composite (UHPFRCC) for impact and blast protection of structures. *Construction and Building Materials* **95**: 667–674.
34. Paschalis S and Lampropoulos A (2017) Fiber content and curing time effect on the tensile characteristics of ultra high performance fiber reinforced concrete. *Structural Concrete* **18**: 577–588.

35. Hannawi K, Bian H, Prince-Agbodjan W and Raghavan B (2016) Effect of different types of fibers on the microstructure and the mechanical behavior of ultra-high performance fiber-reinforced concretes. *Composites Part B: Engineering* **86**: 214–220.
36. Abbas S, Soliman AM and Nehdi ML (2015) Exploring mechanical and durability properties of ultra-high performance concrete incorporating various steel fiber lengths and dosages. *Construction and Building Materials* **75**: 429–441.
37. Gesoglu M, Güneyisi E, Muhyaddin GF and Asaad DS (2016) Strain hardening ultrahigh performance fiber reinforced cementitious composites: effect of fiber type and concentration. *Composites Part B: Engineering* **103**: 74–83.
38. Kazemi S and Lubell AS (2012) Influence of specimen size and fiber content on mechanical properties of ultra-high-performance fiber-reinforced concrete. *ACI Materials Journal* **109**: 675–684.
39. Wu Z, Shi C, He W and Wu L (2016) Effects of steel fiber content and shape on mechanical properties of ultra high performance concrete. *Construction and Building Materials* **103**: 8–14.
40. Wu Z, Khayat KH and Shi C (2018) How do fiber shape and matrix composition affect fiber pullout behavior and flexural properties of UHPC? *Cement and Concrete Composites* **90**: 193–201.
41. Yoo DY, Kang ST and Yoon YS (2016) Enhancing the flexural performance of ultrahigh-performance concrete using long steel fibers. *Composite Structures* **147**: 220–230.
42. Mahmud G, Yang Z and Hassan A (2013) Experimental and numerical studies of size effects of ultra high performance steel fibre reinforced concrete (UHPFRC) beams. *Construction and Building Materials* **48**: 1027–1034.
43. An M, Zhang L and Yi Q (2008) Size effect on compressive strength of reactive powder concrete. *Journal of China University of Mining and Technology* **18**: 279–282.
44. Awinda K, Chen J and Barnett S (2015) Investigating geometrical size effect on the flexural strength of the ultra high performance fibre reinforced concrete using the cohesive crack model. *Construction and Building Materials* **105**: 123–131.
45. Hassan AMT, Jones SW and Mahmud GH (2012) Experimental test methods to determine the uniaxial tensile and compressive behaviour of ultra high performance fibre reinforced concrete (UHPFRC). *Construction and Building Materials* **37**: 874–882.
46. Bastien-Masse M and Brühwiler E (2016) Contribution of R-UHPFRC strengthening layers to the shear resistance of RC elements. *Structural Engineering International* **4**: 365–374.
47. Bastien-Masse M and Brühwiler E (2016) Experimental investigation on punching resistance of R-UHPFRC–RC composite slabs. *Materials and Structures* **49**: 1573–1590.
48. Bastien-Masse M and Brühwiler E (2016) Composite model for predicting the punching resistance of R-UHPFRC–RC composite slabs. *Engineering Structures* **117**: 603–616.
49. Benson SDP and Karihaloo BL (2005) CARDIFRC – development and mechanical properties. Part III: uniaxial tensile response and other mechanical properties. *Magazine of Concrete Research* **57**: 433–443.
50. Japan Society of Civil Engineering (1984) *Part 3-2: Method of Tests for Steel Fiber Reinforced Concrete.* Concrete Library of Japan, Tokyo, Japan.
51. Al-Majidi MH, Lampropoulos A, Cundy A, Tsioulou O and Alrekabi S (2018) Fibre reinforced geopolymer versus conventional reinforced concrete layers for the structural strengthening of RC beams. *Proceedings of the IABSE's 40th Symposium on 'Tomorrow's Megastructures', Nantes, France.* International Association for Bridge and Structural Engineering (IABSE).
52. Cheon H and MacAlevey N (2000) Experimental behaviour of jacketed reinforced concrete beams. *Journal of Structural Engineering, ASCE* **126**: 692–699.

53. Kobayashi K and Rokugo K (2013) Mechanical performance of corroded RC member repaired by HPFRCC patching. *Construction and Building Materials* **39**: 139–147.
54. Rajamane N, Nataraja M, Lakshmanan N and Dattatreya J (2011) Rapid chloride permeability test on geopolymer and Portland cement. *Indian Concrete Journal* 21–26.
55. Ahmad S (2009) Techniques for inducing accelerated corrosion of steel in concrete. *Arabian Journal for Science and Engineering* **34**: 95–104.
56. BS 8110-1 (1997) Structural use of concrete. Part 1: Code of Practice for Design and construction. BSI, London, UK.
57. Lampropoulos A, Tasji A and Tsioulou O (2018) Structural upgrade of deficient unreinforced masonry structures using ultra high performance fibre reinforced concrete. *Proceedings of the IABSE's 40th Symposium on 'Tomorrow's Megastructures', Nantes, France*. International Association for Bridge and Structural Engineering (IABSE).
58. Lampropoulos A, Maisuria N and Tsioulou O (2019) Experimental investigation of the behaviour of unreinforced masonry strengthened with UHPFRC. *Proceedings of the 20th Congress of IABSE: The Evolving Metropolis, New York, NY, USA*. International Association for Bridge and Structural Engineering (IABSE).
59. Abrams D, Smith T, Lynch J and Franklin S (2007) Effectiveness of rehabilitation on seismic behaviour of masonry piers. *Journal of Structural Engineering, ASCE* **133(1)**: 32–44.
60. Galati N, Tumialan G and Nanni A (2006) Strengthening with FRP bars of URM walls subject to out-of-plane loads. *Construction and Building Materials* **20(1–2)**: 101–110.
61. Mosallam A (2007) Out-of-plane flexural behavior of unreinforced red brick walls strengthened with FRP composites. *Composites: Part B* **38**: 559–574.
62. Roca P and Araiza G (2010) Shear response of brick masonry small assemblages strengthened with bonded FRP laminates for in-plane reinforcement. *Construction and Building Materials* **24**: 1372–1384.
63. Lampropoulos A, Tsioulou O, Paschalis S and Dritsos S (2016) Strengthening of unreinforced masonry structures using ultra high performance fibre reinforced concrete (UHPFRC). *Proceedings of the 19th IABSE Congress: Challenges in Design and Construction of an Innovative and Sustainable Built Environment, Stockholm, Sweden*. International Association for Bridge and Structural Engineering (IABSE).
64. Lampropoulos A, Tsioulou O, Paschalis S and Dritsos S (2017) Strengthened unreinforced masonry (URM) structures with ultra high performance fibre reinforced (UHPFRC) layers under axial in-plane and horizontal out-of-plane loading. *Proceedings of the 39th IABSE Symposium: Engineering the Future, Vancouver, Canada*. International Association for Bridge and Structural Engineering (IABSE).

Fibre-reinforced Concretes for High-performance Structures: Building a more sustainable future

Lampropoulos A
ISBN 978-0-7277-6556-7
https://doi.org/10.1680/fchs.65567.171

Chapter 5
Conclusions

In this book, various types of fibre-reinforced concretes were examined, including 'conventional' FRC, UHPFRC with superior mechanical characteristics and FRGC without cement, which aim towards sustainability in construction.

An overview of the main materials required for each type of FRC was presented in the first three chapters together with proposed mix design procedures. Critical evaluation of the mechanical and durability properties of the examined materials was conducted and the importance of the effect of the mix composition, fibre type and fibre volume fraction was highlighted. More specifically, it was found that the addition of fibres significantly enhances the performance of the concrete in tension, which offers improved strength, ductility and energy absorption. Strain-hardening properties are of high importance for load-bearing structural elements as they allow further increment of the tensile strength of the material after its initial cracking. This can be achieved using a high percentage of steel fibres (2% and above by volume of concrete) in combination with a suitable cementitious matrix with high strength characteristics, and this is the main principle for the development of UHPFRC with superior mechanical characteristics. UHPFRC can be used in applications such as thin structural elements or where resistance to extreme loading conditions is required. Recent applications include the use of UHPFRC for the structural strengthening of existing structures, which provides significant benefits in terms of load-bearing capacity and durability. The need for the use of sustainable materials in construction has led to the development of new concrete types, and geopolymer cement-free concretes are the main alternative to Portland cement concrete. In this case, the replacement of cement with waste materials has the potential to lead to a significant reduction of carbon dioxide emissions linked to cement production. Fibre-reinforced geopolymer concretes (FRGC) with high strength characteristics can be successfully used in structural elements. Important parameters for the development of the required strength characteristics include the selection of the appropriate raw materials, the use of suitable activators, appropriate curing conditions and the selection of suitable fibre type and volume fraction. Recent studies have also highlighted the effectiveness of using FRGC for the upgrade of existing structural elements in addition to the corrosion protection of steel bars.

In the fourth chapter, two selected projects/case studies were presented for each of the three examined concrete types (FRC, UHPFRC and FRGC). In the case of FRC, the first project focuses on a research project in which demonstration pavements were constructed using recycled fibres from post-consumer car tyres. Dry mixes were developed and the method of roller compaction was used for the construction of the pavements. The proposed method has significant benefits in terms of cost and energy reduction and offers enhanced durability and reduced maintenance requirements. The second SFRC case study focuses on the design and construction of the Lee Tunnel which is the largest diameter and deepest tunnel in London. The innovative design of the tunnel led to a saving of 1500 tonnes of steel, reduced construction time and significantly enhanced durability. The construction of a cast in situ SFRC secondary lining offers enhanced water resistance and an improved smooth and durable surface. For UHPFRC, two case studies on bridges were presented. The first focuses on the construction of a footbridge where lightweight and durable UHPFRC precast panels were used for the construction with a particular architectural design. The second project focuses on the replacement of an existing deteriorated bridge deck surface using a UHPFRC overlay. The importance of the selection of an appropriate mix design and construction process for the prevention of shrinkage cracking and the enhanced bonding between the overlay and the existing substrate were highlighted. Two case studies using geopolymer concrete were presented. The first focuses on bridge construction using glass-fibre-reinforced polymer (GFRP) bar reinforcement, which gives exceptional mechanical characteristics and durability. This project shows the potential for future developments in this field that aim towards enhanced durability and structural performance. The second of these projects focuses on the use of geopolymer concrete for multi-storey structures for which the use of geopolymer precast segments was found to be the optimum solution after considering other construction types and materials. Finally, in the last part of Chapter 4, the use of high-performance fibre-reinforced concretes (including UHPFRC and FRGC) for the structural strengthening of existing structural elements was examined. Extensive experimental work was conducted as part of research studies and the key parameters for the effectiveness of the use of these materials, such as the fibre orientation, the thickness of the strengthening layers and the presence of additional steel bar reinforcement were examined. The results highlight the efficiency of the use of novel high-performance fibre-reinforced concretes for the structural upgrade of existing structures in addition to the significant benefits linked to durability enhancement.

Lampropoulos A
ISBN 978-0-7277-6556-7
https://doi.org/10.1680/fchs.65567.173

Index

Page numbers in *italics* refer to illustrations.

Printed in the USA
CPSIA information can be obtained
at www.ICGtesting.com
JSHW012041070324
58805JS00005B/18

9 780727 765567